# *BTU Buddy Notebook*

## William (Bill) M. Johnson

DELMAR
CENGAGE Learning™

Australia • Brazil • Japan • Korea • Mexico • Singapore • Spain • United Kingdom • United States

**DELMAR**
CENGAGE Learning

**BTU Buddy Notebook**
**William (Bill) M. Johnson**

Vice President, Career and Professional
Editorial: Dave Garza

Director of Learning Solutions: Sandy Clark

Senior Acquisitions Editor: James Devoe

Managing Editor: Larry Main

Senior Product Manager: John Fisher

Editorial Assistant: Thomas Best

Vice President, Career and Professional
Marketing: Jennifer McAvey

Marketing Director: Deborah S. Yarnell

Marketing Manager: Jimmy Stephens

Marketing Coordinator: Mark Pierro

Production Director: Wendy Troeger

Production Manager: Mark Bernard

Content Project Manager: David Plagenza

Art Director: Bethany Casey

Technology Project Manager:
Christopher Catalina

Production Technology Analyst:
Thomas Stover

For product information and technology assistance, contact us at
**Cengage Learning Customer & Sales Support, 1-800-354-9706**
For permission to use material from this text or product,
submit all requests online at **www.cengage.com/permissions**
Further permissions questions can be emailed to
**permissionrequest@cengage.com**

Library of Congress Control Number: 2008926659

ISBN-13: 978-1-4354-2583-5

ISBN-10: 1-4354-2583-9

**Delmar**
Executive Woods
5 Maxwell Drive
Clifton Park, NY 12065
USA

Cengage Learning is a leading provider of customized learning solutions with office locations around the globe, including Singapore, the United Kingdom, Australia, Mexico, Brazil, and Japan. Locate your local office at
**www.cengage.com/global**

Cengage Learning products are represented in Canada by
Nelson Education, Ltd.

To learn more about Delmar, visit **www.cengage.com/delmar**

Purchase any of our products at your local bookstore or at our preferred online store **www.cengagebrain.com**

Printed in China
2 3 4 5 6 7 16 15 14 13 12

# Table of Contents

## HEATING SEASON

# Introduction

*Introducing "BTU Buddy," Who Says, "Take Some Time to Stop and Think."*

Bob is a service technician who is well-trained and nationally certified. Sometimes, however, he suffers from the same confusion that all technicians occasionally do: The facts that he gathers may or may not point to the obvious cause of the problem. But Bob has something that no one else has. Bob recalls his long-time HVACR mentor and imagines being accompanied by him as "BTU Buddy," someone who reminds him to take time to stop and think before rushing to judgment, helping to keep him on the right track, even with facts that are confusing.

Read on as Bob and BTU Buddy move through a series of residential service calls.

## ABOUT THE AUTHOR

*Bill Johnson*

Bill Johnson graduated from Southern Polytechnic Institute with an associate's degree in Gas Fuel Technology and Refrigeration. He owned and operated an air-conditioning, heating, and refrigeration business for 10 years. He has unlimited licenses for North Carolina in heating, air-conditioning, and refrigeration. Mr. Johnson taught heating, air-conditioning, and refrigeration installation, service, and design for 15 years at Central Piedmont Community College in Charlotte, North Carolina, and was instrumental in standardizing the heating, air-conditioning, and refrigeration curriculum for the state community college system. He is a member of Refrigeration Service Engineers Society (RSES). He is co-author of the successful textbook *Refrigeration and Air Conditioning Technology, 6E.*

# Servicing a Frozen Suction Line and Compressor

In this first call, Bob is sent to a job where the owner complains that the air-conditioning unit ran all night but there is no cooling; the house is still warm and humid. When Bob arrives, he finds that the suction line is frozen all the way back to the compressor and that the compressor is a solid ball of ice. This looks like a clear case of low refrigerant charge.

## A CLEAR CASE?

BTU Buddy says, "Ask some questions, Bob." So Bob asks the owner to describe the sequence of events and the owner says, "We left the unit off all day yesterday while we were at work. When we arrived home last night, we had several people over for a cookout. So we turned the unit on and it ran and seemed to cool for a while, but it ran all night and the house isn't cool now. This is the first time we have run the unit this season."

The outdoor ambient temperature is 75°F, and the indoor temperature is 78°F. BTU Buddy says, "The system is not going to perform correctly under these conditions because of the outdoor ambient temperature." Bob decides to install gauges on the unit.

BTU Buddy says, "This may not be a good idea because the charge will be altered when the 6-foot gauge hose is fastened to the liquid line. This system uses an orifice for a metering device and has a critical charge to plus or minus one ounce. A 6-foot gauge line will hold about 0.21 ounces of R-22 per foot of 1/4-inch line. This figures out to 1.26 ounces of liquid refrigerant that will move into the gauge line. This is enough to affect the system charge during the test. If this refrigerant isn't put back into the system, there will be efficiency problems with the unit. If you are going to install gauges, you should use a shut-off control valve so the refrigerant can be charged back into the unit when you disconnect the gauges.

**FIGURE 1-1.** Shown is a short gauge connection. The valve at the left end controls high pressure while connecting to the system.

You could also use a short-gauge hose connector on the liquid line as shown in Figure 1-1."

## INSTALLING GAUGES

Bob installs the short-gauge connector on the liquid line and a regular 6-foot gauge line on the suction line. He knows that the charge will not be affected by the 6-foot gauge line because that line only contains vapor. When the small amount of refrigerant is released from the lines, it will meet the de minimus (minimum loss) requirement of the EPA.

When Bob installs the gauges, Figure 1-2, he is alarmed at first because the suction pressure is 37 psig and the discharge (head) pressure is 165 psig in this R-22 system. The low side of the system is operating at 14°F, which is well below freezing. BTU Buddy says, "The first thing you must do is defrost the coil."

Bob knows that thawing out the coil will take time so he sets the thermostat to "off" and the fan switch to "fan on." This keeps maximum airflow against the coil and after a while it will defrost a hole in the ice and air will flow through the coil. He tells the owner that he is going to another service call, to give the ice time to melt. He tells her to watch for air to flow at the

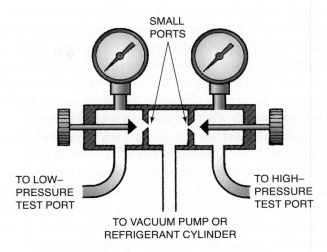

**FIGURE 1-2.** Cutaway of a gauge manifold.

air supply register in the den. When she can feel airflow, she is to call Bob on his cell phone and he'll return.

Bob receives a call in about 2 hours that the air is moving, so he returns. He starts the compressor and observes the pressures and finds the suction pressure to be 55 psig and the head pressure to be 200 psig. The ambient temperature is still 75°F and both pressures are low, still indicating a low charge, so he starts toward his truck to get a cylinder of R-22.

BTU Buddy stops him by saying, "This system is not going to operate correctly because the outdoor temperature is nearly the same as the indoor temperature. The condenser is so efficient that it is condensing too much refrigerant in the condenser. You must remember, this system has a critical charge. There should be a prescribed amount of refrigerant in the condenser, the evaporator, the liquid line, and the suction line. Only then will the system operate correctly. If any amount of refrigerant unbalance occurs, the system will malfunction. In this case, the suction and liquid lines contain the correct charge, as one is vapor and the other is full of liquid. The condenser has too much refrigerant, therefore starving the evaporator. When a system is allowed to run for many hours, the suction pressure and temperature will gradually drop to the point that ice will form—more running time and the coil will freeze. This is exactly what happened last night.

"When the homeowners have a house full of company and the system is not cooling, they have a tendency to turn the thermostat down because they think it will make the unit run faster. When you arrived, the thermostat was set on 60°F and the unit was running continuously. As it got later, the outdoor ambient temperature dropped even further, causing the discharge and suction pressure to drop lower, and a freeze condition occurred.

"Now Bob, block some of the air crossing the condenser and raise the head pressure to about 275 psig, to approximate the head pressure for a 95°F day, and watch what the suction pressure does."

Bob reduces the airflow across the condenser and raises the head pressure to 275 psig, and the suction pressure rises to 70 psig within about 10 minutes. This is close to the design evaporator temperature of 40°F. He then checks the superheat and finds it to be 12°, which is good. BTU Buddy tells Bob, "This is actually simulating a design day for the equipment and can be used to check the charge on most fixed-bore metering devices."

BTU Buddy recaps the call by saying, "It is always good to take some time to think before fastening gauges to a system. Really Bob, you could have left the gauges in the truck, thawed the coil, blocked the condenser airflow until the air leaving the condenser became warm, and felt the suction line to see if it became cold. In doing so, you could have declared the system okay. It isn't good to apply gauges until you really see a need. Service technicians can alter the charge when needlessly installing gauges and may leave a leak behind when removing the gauge lines. Let's agree that a service call should not cause problems, but should solve them."

Bob says, "I agree, it would really be wrong to do more harm than good on a job. I would then be called back to finish the repair. How could I explain that to the customer?"

# Charging an Expansion Valve System after Loss of Charge

Bob receives a call from the company dispatcher that a 10-ton (R-22) air conditioning system at an insurance company is not functioning. There may have been some sort of accident with the condensing unit, which sits outside the office. The evaporator section is just inside the building in an equipment room. When the customer called, she told the dispatcher that the lawn care company reported that they had damaged a pipe and it was spraying a white cloud of something with oil in it all over the place. The dispatcher advised the customer to shut off the unit until a service technician arrived.

When Bob arrives, he goes straight to the condensing unit and discovers the problem: The yard maintenance man cut a small gash in the liquid line between the condensing unit and the building, and refrigerant is still seeping from the gash. Bob decides that he should cut the line and put a coupling in the 5/8-inch OD copper line using high-temperature solder. As he is on his way to the truck to get out the torch and fittings, the insurance company manager comes out and asks how long the repair will take. It is 97°F and the company auditors are in the office going over the books. They are already hot and the manager wants to make them as comfortable as possible. Bob says that it will take at least 3 hours to cut and repair the line; install a filter-drier; leak check, evacuate, and charge the system.

## THE PRESSURE IS ON BOB

BTU Buddy steps in with some advice: "Bob, we all would prefer to go through the repair procedure that you mentioned, but there may be a more efficient way, under the circumstances. Let's go over some other possibilities," he says.

"Here are some things that we know: Refrigerant is still boiling out of the oil in the system and will be for a while, so we know that no air is entering the system. There is no liquid refrigerant in the system, only the vapor refrigerant boiling out of the oil, so there is no chance of recovering any refrigerant without pulling air into the system."

Then he suggests, "Suppose you cut the line and flare both ends and install a flare union. You will be satisfied that no air is in the system, you can then pressure the system with R-22, leak check the flare connection, and charge the system. This way you can have the system back on and charging refrigerant into it within half an hour."

## MAKING THE REPAIR

Bob gets two flare nuts and a coupling for 5/8-inch tubing from his supply and gets started. He cuts the gash out, leaving enough space between the pipes to place the coupling. He then uses his flaring tool, which does not require the tubing to be reamed, to remove any burr left from the cut. The tool has fluted sides on the flare cone and rolls the burr back as the flare is made, as seen in Figure 2-1.

Bob then fastens everything together but the last flare nut connection. He can see vapors being emitted slightly from each fitting, so there is no air entering the system. The flare connection is completed and tightened. When the connections are tightened, Bob installs his gauge manifold and a cylinder of refrigerant, and after purging the gauge lines, allows full tank pressure on the

**FIGURE 2-1.** Flaring tool with a fluted cone that prepares the flare without the need for reaming.

system and leak checks the flare. Since the system has been running fine for several months, it is assumed that the gash in the liquid line is the only leak.

## USING THE PROPER VALVE ARRANGEMENTS

Bob now has a system ready for a charge of refrigerant that is under full tank pressure. He won't be able to push liquid refrigerant into the system because the tank and the system are under the same pressure. He's trying to figure out the quickest way to get the charge into the system.

BTU Buddy says, "Bob, if you start the system it will shut off quickly because of low pressure, so the first thing you should do is place an electrical jumper on the low-pressure control. You may want to use the one that is 3 feet long so you won't accidentally go off and leave it." So Bob jumps out the low-pressure control. BTU Buddy then advises him to shut the king valve on the liquid line and turn the refrigerant cylinder valve to the liquid opening. Then Bob starts the system. The low side gauge begins to drop and refrigerant feeds from the liquid connection on the refrigerant tank into the liquid line. This is a 10-ton system, so it will take several pounds of refrigerant. The liquid line gets very cold as liquid enters it. The system is actually using the refrigerant cylinder as the receiver and will start cooling very quickly.

## BRINGING THE SYSTEM TO THE CORRECT CHARGE

When the system has run for a few minutes, Bob shuts off the refrigerant cylinder liquid valve and opens the king valve on the liquid line. The suction pressure is about 50 psig and holding, so the charge is fairly close to correct. The system will operate at this condition, so Bob removes the jumper wire from the low-pressure control and then starts metering liquid refrigerant very slowly into the suction line. Barely opening the liquid line, he only lets the suction pressure rise to about 10 psig above the operating suction pressure. He lets refrigerant in very slowly until the sight glass entering the expansion valve is clear.

Bob goes inside and checks that the building is beginning to cool down. The manager is very thankful that the system is beginning to cool the office.

Bob starts to pack up his tools and BTU Buddy says, "Hold on, Bob. This job is not finished." Bob says, "The system sight glass is full of liquid and the system is cooling; what else is there to do?" BTU Buddy points out, "The system will run and cool just like you have charged it, but it is not operating at its maximum capacity or efficiency. The only thing you've done up to now is furnish pure liquid to the expansion valve. What about subcooling for full efficiency?"

"Oh yeah," says Bob, "I forgot about that."

Bob goes to his truck and gets a temperature tester and fastens one of the leads to the liquid line, insulating it so the ambient air will not affect it. The ambient temperature is 97°F so the system will be operating at about outside design conditions. The inside is still warm but will be down to about design temperature in a few minutes. Bob remembers that equipment is designed to operate at 95°F dry bulb outside, and 80°F dry bulb with 50% relative humidity inside. Bob also knows that even though the equipment is rated for those conditions, no one is comfortable at 80°F dry bulb; so he and most other technicians use 75°F for their inside operating conditions.

BTU Buddy says, "The system should have about 10–15°F of subcooling to be operating correctly. The system will gain about 1 percent capacity for each degree of subcooling, at no additional cost.

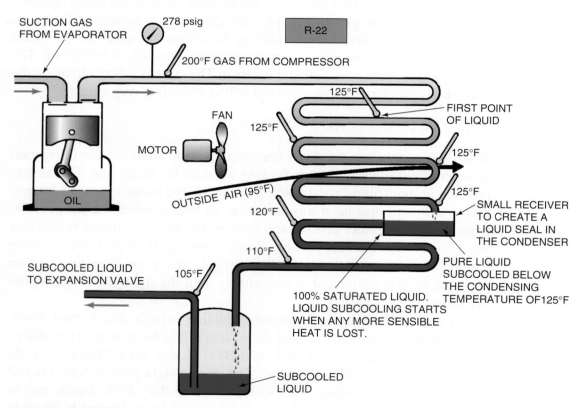

**FIGURE 2-2.** The condenser has a small reservoir that forms a liquid seal between the condensing portion of the condenser and the subcooling circuit.

Bob checks the liquid line temperature and compares it to the condensing temperature on the gauges. The head pressure is 297 psig and the liquid line temperature is 128°F. The system saturation or condensing temperature is about 130°F, so there are 2 degrees of subcooling.

Bob adds refrigerant slowly to the suction line and watches the head pressure and the liquid line temperature. As he adds refrigerant, the liquid line temperature drops. BTU Buddy tells Bob, "To reach a subcooling of 10°F, the liquid line temperature needs to be 120°F. For a subcooling of 15°, the liquid line temperature needs to be 115°F. As the refrigerant is added to a system, the head pressure should remain constant at 297 psig for this application. Once the saturation temperature is reached, it should not change while adding refrigerant until the system is overcharged. Adding refrigerant to obtain subcooling does not affect the condensing circuit because it is a circuit all its own, as shown in Figure 2-2. Since the pressures are the same with just a full sight glass as with a full sight glass and a 15° subcooling charge, the operating cost is the same. The system gains capacity by furnishing subcooled liquid to the expansion valve. The only cost is additional refrigerant."

As Bob is loading his truck, the manager comes out and says, "I can't thank you enough for getting this job done so efficiently. The office is cool again and the auditors' attitudes have improved."

While Bob is driving away, BTU Buddy says, "Being professional only takes a little more time, to learn what to do and use what you know. The difference between a professional and an amateur athlete is that one is dedicated to a career of high earning potential while the other is just there to play around. This is much the same in our business, Bob, and you are a professional because you care."

# Charging a System with a Fixed-Bore Orifice

*Cooling Off a Very Hot Compressor*

Bob gets a service call from the dispatcher about a residence where a very sick woman is living, and the air-conditioning system is not working properly. It is 96°F outside, and the woman's bedroom is upstairs, where the heat tends to build up quickly.

Bob arrives at the job and asks the owner what the sequence of events was. He says that the unit was cooling yesterday, but is not doing well today.

Bob goes to the outdoor unit. The fan is running, but the compressor is not. He shuts off the disconnect switch and removes the compressor compartment door. The compressor is very hot. There is oil in the compressor compartment and he can hear the refrigerant leaking. Bob looks around and finds that the discharge line has been rubbing on the cabinet and there is a very small leak there. Refrigerant is still leaking rapidly, so he decides to recover it.

## RECOVERING THE REFRIGERANT

Bob connects his recovery machine and when the manifold gauge approaches 0 psig, BTU Buddy makes an appearance and asks, "When are you going to stop the recovery machine?" Bob says, "When it gets to about 28 inches of mercury on my manifold gauge."

"That may not be the best idea," says BTU Buddy. "You know you have a leak and if you allow the system to go into a vacuum, air will enter the system." Bob says, "Not much air will get in through that small hole."

BTU Buddy reminds him, "No air should knowingly be allowed into the system. There is a better way. Stop the recovery at 0 psig, repair the leak, and charge the system. No air enters, the repair is made in a timely fashion, and the system is back on-line in a minimum of time." Bob stops the recovery at 0 psig on the manifold gauge as BTU Buddy suggested.

Bob uses 15% silver solder and closes the hole in the discharge line. He gently moves the line away from the cabinet, where it will not rub again. It's time to start the system. Bob tells BTU Buddy, "This system uses R-22 and has an orifice type of metering device, and we don't know what the system charge is supposed to be." BTU Buddy explains, "We can charge the system using the superheat method. The ambient temperature is 96°F and this is an ideal day for charging with that method. We need to know the approximate length of the suction line."

Bob measures the line length to be about 40 feet. The air handler is under the house on the far side from the condensing unit. "Why do we need to know the line length?" asks Bob.

"We really need to be able to check the superheat at the air handler by checking the suction line temperature and the suction pressure at the air handler," says BTU Buddy. "But we cannot readily check the suction pressure there and it's not convenient to check the temperature at the evaporator, so we're going to take these measurements at the condensing unit and approximate the temperature rise in the suction line by taking the temperature at the condensing unit. We can take the pressure at the same place as the temperature and get an accurate superheat reading. The reading is just at the condensing unit, not the evaporator.

"The superheat in the suction line is greater at the condensing unit than at the evaporator because it picks up heat all along the way, so we will make some assumptions in order to charge the unit as close to the factory charge as possible. We could use three different line lengths for determining what superheat to use: up to 10 feet, 10 feet to 30 feet, and 30 feet to 50 feet. For line lengths up to 10 feet, we'll use 10°F of superheat. For line lengths of over 10 feet to 30 feet, we'll use 10 to 15°F of superheat, and for line lengths of 30 feet to 50 feet, we would use 15 to 18°F of superheat. This is probably as accurately as we can charge a unit with an unknown charge in the field."

## STARTING THE UNIT

Bob has made the repair and is ready to start the unit. He connects the cylinder of R-22 to the system to charge liquid refrigerant into the liquid line. The system pressure is 0 psig, so he opens the manifold gauge to allow pure liquid into the liquid line connection. The liquid refrigerant will flow both ways, toward the evaporator and toward the condenser. When the liquid stops entering, he shuts off the liquid and prepares to charge into the suction line. He turns on the system and the outdoor fan starts, but the compressor doesn't start. Bob touches the compressor and it's still hot. Bob says to BTU Buddy, "I hope the compressor isn't burned. It's very hot."

BTU Buddy tells Bob, "It's not likely unless it burned just before we arrived. We could find out for sure by turning off the power and testing the compressor electrically, or we can cool the compressor and try it again. There are three ways we can cool the compressor:

1. Wait until tomorrow. But remember, the sick woman needs relief.
2. Leave the compressor compartment door open and run the fan. This only works when the compressor compartment is open to the outdoor fan air-flow, and it may take several hours. Some units have isolated compressor compartments and air will not flow over the compressor.
3. Cool the compressor with water. The compressor is not touching the compressor shell; it is suspended in the vapor space inside. This takes a few minutes, but it's the quickest of the three ways.

Since it's a long drive back to town and back out here tomorrow, and the customer really needs the system to be working soon, let's cool it with water."

Bob asks, "How do we do that without danger of electrical shock?" BTU Buddy tells Bob, "Shut off the power to the unit and lock the box so no one else can turn it on. You have a personal lock and key. Now, get a water hose and lay the nozzle on top of the compressor and let water trickle over the compressor dome, as depicted in Figure 3-1. This will take about 15 minutes,

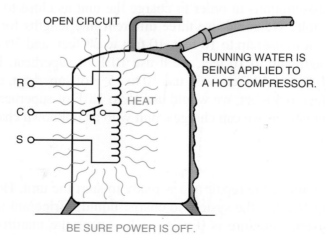

FIGURE 3-1. Cooling a compressor with water. The unit is shut off and the electrical panel is locked. Water is allowed to trickle over the compressor to cool it. It normally takes about 15 minutes due to the compressor itself being suspended in vapor refrigerant within the compressor shell.

so let's use the time to the customer's advantage and change the air filters and oil the indoor fan motor."

Bob returns to the condensing unit in about 25 minutes and turns off the water. BTU Buddy says, "Now, let the water around the unit drain away. You will find that this is no more dangerous than working on a unit in the rain, which you often have to do."

Bob turns on the unit, and the compressor starts up this time. BTU Buddy tells Bob, "Meter liquid refrigerant slowly into the suction line, keeping the pressure about 10 psig above the actual suction pressure. That motor is still hot enough to vaporize any liquid that reaches it."

Bob notices that the actual suction pressure is 50 psig, so he allows the entering liquid to raise the suction pressure to about 60 psig, as illustrated in Figure 3-2. Bob fastens his temperature tester to the suction line and insulates it from the ambient air so he will get a true suction line gas temperature.

Bob shuts off the refrigerant flow into the system for a superheat check. The suction line temperature is 70°F and the suction pressure is 61 psig,

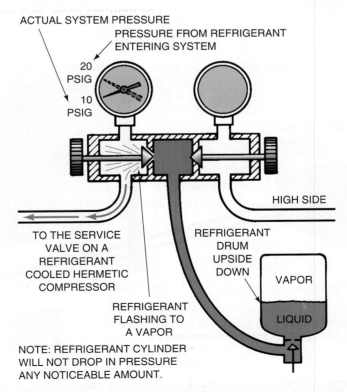

**FIGURE 3-2.** A manifold gauge is used to meter the liquid refrigerant to a vapor by not allowing the pressure to exceed 10 psig above the running suction pressure while adding refrigerant.

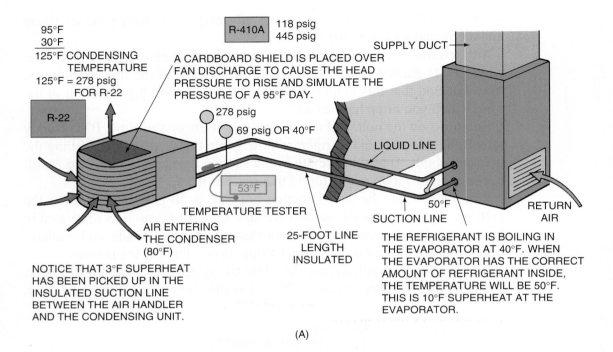

95°F
30°F
125°F CONDENSING
TEMPERATURE
125°F = 278 psig
FOR R-22

R-22

R-410A
118 psig
445 psig

SUPPLY DUCT

A CARDBOARD SHIELD IS PLACED OVER
FAN DISCHARGE TO CAUSE THE HEAD
PRESSURE TO RISE AND SIMULATE THE
PRESSURE OF A 95°F DAY.

278 psig

69 psig OR 40°F

LIQUID LINE

53°F

TEMPERATURE TESTER

AIR ENTERING
THE CONDENSER
(80°F)

25-FOOT LINE
LENGTH
INSULATED

50°F

SUCTION LINE

RETURN
AIR

NOTICE THAT 3°F SUPERHEAT
HAS BEEN PICKED UP IN THE
INSULATED SUCTION LINE
BETWEEN THE AIR HANDLER
AND THE CONDENSING UNIT.

THE REFRIGERANT IS BOILING IN
THE EVAPORATOR AT 40°F. WHEN
THE EVAPORATOR HAS THE CORRECT
AMOUNT OF REFRIGERANT INSIDE,
THE TEMPERATURE WILL BE 50°F.
THIS IS 10°F SUPERHEAT AT THE
EVAPORATOR.

(A)

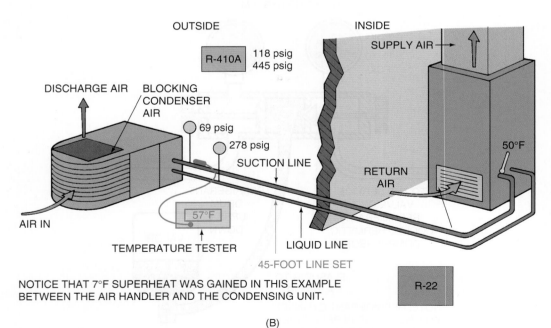

OUTSIDE

INSIDE

R-410A
118 psig
445 psig

SUPPLY AIR

DISCHARGE AIR   BLOCKING
CONDENSER
AIR

69 psig

278 psig

SUCTION LINE

50°F

RETURN
AIR

AIR IN

57°F

TEMPERATURE TESTER

LIQUID LINE

45-FOOT LINE SET

NOTICE THAT 7°F SUPERHEAT WAS GAINED IN THIS EXAMPLE
BETWEEN THE AIR HANDLER AND THE CONDENSING UNIT.

R-22

(B)

**FIGURE 3-3.** System charged by raising the discharge pressure to simulate a 95°F day. This is a standard efficiency
system that is charged to the correct charge for a 45-foot line set, very similar to the system in this service call.

which corresponds to about 35°F—refrigerant boiling temperature. The superheat is 35°F (70°F suction line temperature minus 35°F boiling temperature taken from the low side-gauge pressure reading converted to temperature), so there is not enough refrigerant in the system yet. Bob starts to add liquid again.

BTU Buddy tells Bob, "I think it would be a good idea to change over to vapor before you overcharge the unit. It is much easier to charge a unit using superheat by adding vapor than to overcharge it and recover refrigerant to the correct level of superheat."

Bob asks, "How in the world do you get enough experience to know all of these things?" BTU Buddy says, "Bob, experience cannot be bought or learned from books alone, you have to get out and make mistakes. Most people don't learn from their mistakes, they just keep making the same ones over and over. It reminds me of a technician who bragged that he had 10 years experience. After working with him awhile, I discovered that he had one year of experience, 10 times over. He never learned anything after the first year. The education of a good technician never ends. The technician just stays interested in the subject and learns forever."

Bob keeps charging the system until the suction pressure is 69 psig, which corresponds to approximately 40°F. The suction line temperature was reading 57°F, so the superheat is 17°F, which is very good, Figure 3-3. He then shuts the valve on his liquid-line gauge hose and pulls the liquid refrigerant from the gauge line into the low side of the system, minimizing the amount of refrigerant lost, as all that was left was vapor at the low side pressure. A final check of superheat shows it to be 16°F, well within the range for a correct charge. The final amount of refrigerant from the liquid line lowered the superheat a small amount.

"Well, Bob," says BTU Buddy, "you learned a valuable lesson from this call and you got the system back on in record time. The homeowner always appreciates good service."

# Cleaning a Very Dirty Air-Cooled Condenser

The company dispatcher calls Bob and asks him to take a service call on his way home late in the afternoon. The homeowner is complaining of an overly hot house and says her air-conditioning unit is running all the time.

Bob arrives at the home about 5:00 p.m. to discover that the temperature inside the house is 80°F while the outside temperature is 99°F. This is the first really hot day in a region known for heat. Bob asks the owner what she knows about the way the system is performing. She tells Bob that the unit has run all day and doesn't seem to be getting anywhere.

Bob goes to the back of the house and discovers the condensing unit is running—at least the fan is running. The compressor is not running, so he removes the compressor compartment door. The compressor is too hot to keep his hand on. Bob looks over the unit and quickly makes a decision that the condenser coil is extremely dirty and must be cleaned to operate correctly. While the compressor is cooling down, Bob prepares to clean the condenser with a water hose.

About this time, BTU Buddy appears and asks, "What procedure are you about to perform?" Bob says, "I am going to use water to wash this condenser from the outside—that will save time." BTU Buddy suggests, "Bob, if you only use water and wash from the outside inward, you are going to be making two mistakes. One, water alone will not clean a condenser, and two, you need to backwash the condenser in the opposite direction of the airflow for proper cleaning."

"Why the opposite the direction of the airflow?" asks Bob. BTU Buddy explains, "The heaviest dirt will be where the air enters the coil. If you wash the coil in the direction of the airflow, you may drive the dirt deeper into the core of the coil, which has several rows. By backwashing, you push the dirt outward in the direction in which it entered the coil."

## CLEANING THE COIL

Bob gets a chemical detergent that is approved for cleaning a copper coil with aluminum fins; one that will not damage either metal. BTU Buddy tells Bob, "Some of the very harsh cleaners that people have used may cause deterioration of the connection between the aluminum fins and the copper tube in the coil. This connection is vital for good heat transfer."

Bob turns off the power to the unit. BTU Buddy reminds Bob, "You should lock the disconnect box and keep the key in your pocket. You could get a phone call and walk out to the truck for a moment and someone, a kid or someone else, might turn the disconnect back on. Then, when you spray water on the unit, you could become an electrical circuit to ground. It should become a habit to always lock the disconnect box while working on the electrical system or cleaning the unit with water."

Bob says, "It's hard to remember all the correct steps for a service call." BTU Buddy says, "When you realize that many of these steps are safety steps and can be life saving, you will remember. It seems that your memory is often in proportion to your responsibility. The more kids you have, the better your memory will become for safety details."

Bob removes the top from the condensing unit and discovers that the wiring harness to the fan motor mounted in the top is long enough to lay the fan down beside the unit. The manufacturer did a good job on this. If the wiring harness were too short, the fan would have to be disconnected to move it to the side; then it would have to be reconnected later. Bob uses a sprayer, similar to a garden sprayer, to apply approved cleaning detergent to the condenser. He starts spraying from the inside, then goes to the outside, and saturates the coil with cleaner.

BTU Buddy says, "This coil is going to need to soak for about 15 minutes to let the detergent do its work. Why don't you use this time to change the air filters, and oil the indoor fan motor and the outdoor fan motor. The customer will get maximum value for your time, and believe me, they will notice that—particularly if you tell them you have extra time and you are going to use it to their advantage. I believe you should always give added value and inform the customer. They will ask for you to be their service connection and be loyal to your company for the long term."

The coil soaked for about 20 minutes and Bob returned. He gave the coil another coating of detergent and checked the compressor contactor while the detergent was soaking in. Then he spread plastic over the control panel and fan motor to protect them from water spray. He took the water hose with a spray nozzle and began to spray through the coil from the inside out.

Bob could not believe the trash that came out of the coil. It was full of grass clippings at the bottom.

"I wonder what this is all about?" asks Bob. BTU Buddy notes, "Look around and observe how the yard is cut. Look at the grass clippings on the side of the house up close to the condensing unit." Bob says, "It looks like the yard person isn't using a grass catcher and is mowing with the grass discharge aimed at the house. When it goes by the unit, the grass clippings are hitting it, and if the unit is running, it will suck them into the coil, particularly at the bottom."

"Good observation," says BTU Buddy. Bob goes and gets the homeowner's wife and shows her what has been happening. About that time the husband comes home from work and Bob shows him. He says, "I cut the grass and never thought of what it was doing. From now on, I'll either use a grass catcher or reverse the direction that I cut the grass in to blow the clippings away from the unit."

## DETERMINING WHETHER THE CHARGE IS CORRECT

Bob puts the top of the unit back on and starts for the truck. BTU Buddy asks, "Where are you going, Bob?" Bob says, "I'm going for my gauges and ammeter to observe with when I start the system." BTU Buddy says, "Just get your ammeter, I'm going to show you a trick." Bob returns with a clamp-on ammeter and clamps it on the common wire going to the compressor, as shown in Figure 4-1. Bob says, "The run load amperage is 23 amps."

BTU Buddy tells Bob, "The question is whether the unit has enough refrigerant. If we put gauges on the system, we may alter the charge when

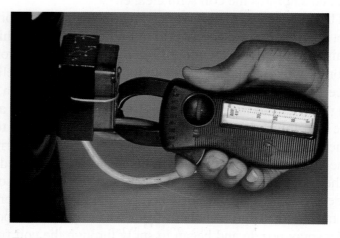

**FIGURE 4-1.** The clamp-on ammeter is used to measure the common wire to the compressor.

it's already charged correctly. So let's try something else. If we put gauges on the unit, the gauges will not read anywhere near what you would expect them to read. The entering evaporator air is 85°F and we have no idea what the humidity is. The increased humidity will add more load to the evaporator. We know that the indoor coil has a tremendous load on it. The condenser is going to have an extra load from the indoor coil, plus it is 99°F, which is going to impose an even greater load on the system compressor. We know that when we arrived the compressor was hot. We believe it was overloaded because of the dirty condenser. Now, here is what we're going to do. Put your hand on the suction line and get a good grip on it. Now when you reach over with the other hand and start the unit, tell me what you feel in the first few seconds of operation." Bob starts the unit and in a few seconds, a puzzled look comes over his face. He tells BTU Buddy, "The suction line felt really cool for a few seconds, and then it began to warm. What's going on here?"

BTU Buddy explains, "With the system off, some of the liquid refrigerant is in the evaporator and some in the condenser. When you start the unit, if the charge is correct in the system, some of the liquid from the evaporator will move down the suction line and you will be able to feel it for a few seconds. If you walk over and turn the unit on and walk back to the unit, you may miss this event. This tells us that the unit most likely has enough refrigerant. Because gauge readings would be hard to interpret under the overloaded conditions, we will assume the charge is correct. This can be further proved by the compressor amperage. What is the amperage running?"

Bob looks at the ammeter and it is showing 23 amps. Because the compressor is fully loaded, this indicates that the compressor is working and it is likely that the charge is correct in the system. BTU Buddy says, "You need to explain to the owners that the conditions in the house will not return to normal until late tonight or maybe even morning. The unit will have to lower the humidity before the temperature in the house begins to drop. They'll notice an abnormal amount of water coming out of the condensate drain during this time."

Bob explains this to the homeowners and packs up to leave. BTU Buddy adds one more thought, "Bob, you have given the customer extra service and really paid attention to their needs. I suspect you will be on their list of people that they want to do their work from now on." Bob tells BTU Buddy, "It takes a lot of concentration and knowledge to become efficient in this business. I wish everybody had a BTU Buddy to consult with."

# A Conversation about Vacuum Procedures

Bob stops at a diner for lunch and has something on his mind. He has been to school where all of the terms of pressure and vacuum were discussed, but he doesn't have a firm grasp on the meanings of the terms. He wants to better understand the evacuation of a system. At about this time, BTU Buddy appears to clear up Bob's questions.

"Think back about how an evacuation works, Bob, and we'll discuss some terms," says BTU Buddy. "The vacuum pump removes vapor from a system. Some technicians think it will remove solids and acids, but it only removes vapor. If acid is in the form of smoke from a motor burn, it will remove that. A vacuum pump is a 'vapor pump' just like a compressor, except it is capable of pumping in very low pressure ranges."

He explains, "Vapor molecules in a vessel travel in a straight line and bounce around the vessel. If there is an opening in the vessel, the molecules will begin to move out of the opening in proportion to the pressure difference from the inside of the vessel to the outside, as shown in Figure 5-1. There are many more molecules at a higher pressure. If the vessel is under 20 psig of pressure and atmospheric pressure is 0 psig, there is a 20 psi incentive for the vapor to move out. There are a lot of vapor molecules at 20 psig. When the vessel pressure is at 0 psig, there is not much incentive for the molecules to move out of the vessel. The vacuum pump provides the incentive by creating a low pressure area for the molecules to move to. The vacuum pump then exhausts the molecules out into the atmosphere."

Bob tells BTU Buddy, "I remember all of this from school, but I can't say that I really understood it."

BTU Buddy says, "To really understand what this low pressure is, you must understand the units of that pressure. It's much like going to a foreign country; you must understand the currency or you never know where you stand."

**FIGURE 5-1.** Molecules of gas in a container move around. If there is an opening, the molecules will begin to move out of the opening in proportion to the pressure difference from the inside to the outside.

BTU Buddy continues, "All pressure readings relate to the force of a standing column of water, since that was probably the first kind of measuring device used for pressure. For example, a vessel with a pressure in it of 1 psig will support a column of water 2.31 feet (27.7 inches) high, Figure 5-2. Inches of water column are still used for measuring gas pressures and air pressures in duct systems, as these pressures are very low. The atmosphere will support a column of water about 34 feet high. This is the same as 14.696 psi, often rounded to 15 psi.

"The weatherman talks about inches of mercury (Hg) to express atmospheric pressure. We often use millimeters of mercury to express pressures in our field. The weatherman would tell you that the atmosphere will support a column of Hg that is 29.92 inches high, as shown in Figure 5-3. The 29.92 inches is often referred to as 30 inches. It is an equal pressure to 33.94 feet of water and is the measure of sea level atmosphere on a standard pressure day. Since these two pressures are equal, and used by everyone in our field, you must be able to convert from one to the other. In reality, 1 psi is equal to 2.036 inches of Hg. This can be verified by dividing 29.92 by 14.696, (29.92 ÷ 14.696 = 2.0359, rounded to 2.036). The 2.036 is called a conversion factor or unit."

Bob breaks in and says, "That has always been confusing to me! To make it harder, technicians talk about both in the same conversation. I don't understand why we have to know all of this."

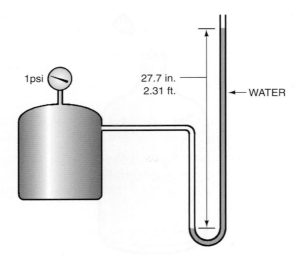

**FIGURE 5-2.** This vessel is registering a pressure of 1 pound per square inch gauge and it will support a column of water 2.31 feet high.

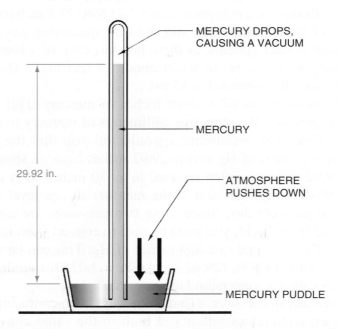

**FIGURE 5-3.** This is a basic mercury (Hg) barometer. Standard atmospheric pressure will support a column of Hg 29.92 inches high.

BTU Buddy says, "For the very reasons you just said; it is the language of HVACR."

Bob asks, "But how are we going to use it?"

BTU Buddy says, "When you buy your instruments and vacuum pump for removing the atmosphere, all of these terms come up and more. For example, we have not even talked about the terms used to measure a deep vacuum. Remember, the atmosphere will support a column of Hg 29.92 inches high.

"In the U.S., we express positive pressures in pounds per square inch gauge. Our vacuum gauge registers pressures below atmospheric and has a vacuum scale expressed in inches of Hg vacuum. This scale is needed because it is divided into smaller increments. We talk in terms of millimeters of Hg down into the very low vacuums that must be obtained to properly evacuate a system."

Bob says, "Yes, I know the term but can't say that I can explain it to anyone else."

BTU Buddy notes, "When it comes to deep vacuums, manufacturers use even more terms that you should be familiar with. Let's say it this way: The atmosphere will also support a column of Hg 760 millimeters high. Just remember, all of the terms for measuring a column have been simple linear measurements—feet, inches, and millimeters. These terms can easily be converted from one to another. Before we put together a conversion table, we must discuss one more important term, microns. A micron is equal to 1/1000 of a millimeter. A millimeter is about 1/2 the thickness of a dime.

"The object of reducing the pressure in a system to a deep vacuum, known as pulling a vacuum, is to remove vapor from the system. This is measured in microns. Many manufacturers will say in their specifications that they want you to pull a vacuum to 250 or 500 microns and hold that pressure for a time span to prove there are no foreign vapors in the system. This would include water vapor. A pressure reading of 500 microns is actually 1/2 of a millimeter and cannot be readily detected by the naked eye with a column of Hg."

BTU Buddy then says, "You must realize by now that your manifold gauge will not measure pressures this low. That's why your company furnishes you with an electronic gauge to measure vacuum, which is calibrated to read in microns. Your manifold gauge could better be called an indicator. It is not a precise measuring instrument."

Bob says, "You know, I've used that micron gauge a lot, but never really understood what it was all about."

BTU Buddy says, "Let's construct a simple chart that explains the pressure conversions."

1 inch = 254 millimeters
1 millimeter = 1,000 microns
1 psi = 2.036 inches of Hg
760 millimeters = 29.92 inches of Hg = 14.696 psig = 33.9 feet of water

BTU Buddy remarks, "When you understand the relationships of these facts, you can talk intelligently to anyone about pressures and vacuums."

Figure 5-4 is a table of all of these relationships along with the saturation points for boiling water.

BTU Buddy tells Bob, "A good vacuum pump can lower the pressure to 0.1 micron. A micron is 1/1000 of a millimeter of pressure above absolute 0 pressure expressed in millimeters Hg (remember, there is also pounds per

| ATMOSPHERIC PRESSURES, ABSOLUTE VALUES | | | | COMPOUND GAGE READING in. Hg VACUUM | SATURATION POINTS of $H_2O$ (BOILING—CONDENSING) °F |
|---|---|---|---|---|---|
| psia | in. Hg | mm Hg | microns | | |
| 14.696 | 29.921 | 759.999 | 759,999 | 00.000 | 212.00 |
| 14.000 | 28.504 | 724.007 | 724,007 | 1.418 | 209.56 |
| 13.000 | 26.468 | 672.292 | 672,292 | 3.454 | 205.88 |
| 12.000 | 24.432 | 620.577 | 620,577 | 5.490 | 201.96 |
| 11.000 | 22.396 | 568.862 | 568,862 | 7.526 | 197.75 |
| 10.000 | 20.360 | 517.147 | 517,147 | 9.617 | 193.21 |
| 9.000 | 18.324 | 465.432 | 465,432 | 11.598 | 188.28 |
| 8.000 | 16.288 | 413.718 | 413,718 | 13.634 | 182.86 |
| 7.000 | 14.252 | 362.003 | 362,003 | 15.670 | 176.85 |
| 6.000 | 12.216 | 310.289 | 310,289 | 17.706 | 170.06 |
| 5.000 | 10.180 | 258.573 | 258,573 | 19.742 | 162.24 |
| 4.000 | 8.144 | 206.859 | 206,859 | 21.778 | 152.97 |
| 3.000 | 6.108 | 155.144 | 155,144 | 23.813 | 141.48 |
| 2.000 | 4.072 | 103.430 | 103,430 | 25.849 | 126.08 |
| 1.000 | 2.036 | 51.715 | 51,715 | 27.885 | 101.74 |
| 0.900 | 1.832 | 46.543 | 46,543 | 28.089 | 98.24 |
| 0.800 | 1.629 | 41.371 | 41,371 | 28.292 | 94.38 |
| 0.700 | 1.425 | 36.200 | 36,200 | 28.496 | 90.08 |
| 0.600 | 1.222 | 31.029 | 31,029 | 28.699 | 85.21 |
| 0.500 | 1.180 | 25.857 | 25,857 | 28.903 | 79.58 |
| 0.400 | 0.814 | 20.686 | 20,686 | 29.107 | 72.86 |
| 0.300 | 0.611 | 15.514 | 15,514 | 29.310 | 64.47 |
| 0.200 | 0.407 | 10.343 | 10,343 | 29.514 | 53.14 |
| 0.100 | 0.204 | 5.171 | 5,171 | 29.717 | 35.00 |
| 0.000 | 0.000 | 0.000 | 0.000 | 29.921 | — |

NOTE: psia × 2.035 966 = in. Hg     psia × 51.715 = mm Hg     psia × 51,715 = microns

**FIGURE 5-4.** This chart shows the common pressures below atmospheric pressure in relationship to each other as well as a pressure temperature saturation chart for water ($H_2O$).

square inch absolute, psia). **When the vacuum pump manufacturers say this, they mean in a clean system with nothing that can boil to a vapor. This will not work in a refrigeration system because we have oil in the compressor and piping that will emit small amounts of vapor. That is why manufacturers use 250 to 500 microns.** With a pressure that is this low in the vacuum pump, the system pressure is extremely low, ensuring that there is no moisture in the system after evacuation."

Bob says, "There are sure a lot of details to evacuation. Can I ever learn them all?"

BTU Buddy says, "The learning curve is long in this business. It may never end because new technologies are introduced every day. That's what makes it an interesting career; you never see an end to the variety of knowledge. A person who is making a career of this field can learn forever. It is a constant educational challenge."

# Removing Water from a Wet Heat Pump System

Bob is scheduled to help finish a job that a new construction crew installed. The installers left the gas line and liquid line to a heat pump at the job site turned up all weekend, and it rained into the tubing. The foreman noticed it and asked for the service department to finish the job.

Bob arrives at the job and sees right away what the problem is. Sure enough, the new construction department has made a mess of this one. The installation crew ran the gas and liquid lines out to the outdoor unit and just turned them up on the ends, intending to plug them before leaving for the weekend, but then forgetting. There are no gutters on the house, and water from the roof just ran into the lines. The lines had a fairly long horizontal run and were all standing full of water. This would have been the end of these systems if they were just connected to the outdoor units and started up. The foreman made a good call in getting the service department involved at this time.

Bob has turned the lines down to drain as much water out as possible, and is standing there wondering what to do next when BTU Buddy appears and says, "Why don't you solder 1/4-inch connectors in the liquid lines first, and purge the lines with nitrogen to push as much water as you can through the system and out the gas lines?"

"Why connect to the liquid line first?" asks Bob.

"Well, the bulk of the freestanding water will be in the large line, the gas line. This way, you can push most of the water out toward the large line. You could actually go up and remove the metering device and really get the best velocity through the line, but it is brazed in the system," says BTU Buddy.

Bob does as BTU Buddy suggests and prepares to purge the system with nitrogen. He finds that the upstairs system has very little water in it. When he turns the pressure on to the downstairs system, water just keeps blowing out. The indoor coil for this system has a lot of water in it, whereas the upstairs unit only has water in the piping.

BTU Buddy says, "Bob, you need to use every advantage you have to evacuate this system to a very deep vacuum. This is a low temperature system in the winter. If there is a drop of moisture circulating in the system, it will eventually get to the metering device in the winter when the system is operating below freezing, and stop the entire system."

Bob asks, "Will a single drop of water really stop the system from functioning?"

"Absolutely," says BTU Buddy. "The industry has always said that a good low-temperature refrigeration technician makes a great air-conditioning technician. The reason is that refrigeration technicians really understand moisture in a system. It just cannot happen. Moisture can actually circulate in an air-conditioning system because it never operates below freezing, but it cannot be tolerated in a low-temperature system. You should never allow moisture in an air-conditioning system for other reasons besides freezing. Moisture will combine with the refrigerant and heat in the compressor and create acid—not a lot of acid—but enough to cause electroplating in the compressor.

"Remember, the elements that must be present for electroplating are acid, electricity, and dissimilar metals. The acid is formed from the moisture, there is electricity from electrical supply in the hermetic compressor, and there is copper and steel in the system. The electroplating normally plates from the soft metal to the hard metal, so the copper is plated to the steel crankshaft and wrist pins. This closes the tolerance between the shaft and the rod and causes it to bind. This is one reason why a compressor may not start."

BTU Buddy continues, "Most of what you remove from a system when you evacuate it is nitrogen from pressurizing the system, or atmosphere that may have small amounts of moisture because the system has been open. These are very easy to remove."

Bob asks, "How much water do you think is still in the downstairs system?"

BTU Buddy says, "It's hard to tell. You really can't get much velocity through the pipes with the metering device still in the system, but we can use some tricks and get it cleaned up."

BTU Buddy explains, "If there is freestanding water in the system, this water must be boiled to a vapor before it can be removed. Water actually creates large volumes of vapor. One pound of water will generate about 867 cubic feet of vapor when boiled (evaporated) at 70°F. If a large vacuum pump is used to dehydrate a system after a water leak, it may boil the water at an even lower temperature, such as 40°F, where it would create 2,444 cubic feet of vapor, as seen in Figure 6-1. If the moisture is boiled too fast, it can actually turn to ice which is really hard to remove. The typical technician working on residential equipment may not ever have to evacuate a flooded system. When large volumes of vapor have to be removed in an evacuation, the correct procedures must be used."

| TEMPERATURE | | SPECIFIC VOLUME OF WATER VAPOR | ABSOLUTE PRESSURE | | |
|---|---|---|---|---|---|
| °C | °F | ft³/lb | lb/in.² | kPa | in. Hg |
| −12.2 | 10 | 9054 | 0.031 | 0.214 | 0.063 |
| −6.7 | 20 | 5657 | 0.050 | 0.345 | 0.103 |
| −1.1 | 30 | 3606 | 0.081 | 0.558 | 0.165 |
| 0.0 | 32 | 3302 | 0.089 | 0.613 | 0.180 |
| 1.1 | 34 | 3059 | 0.096 | 0.661 | 0.195 |
| 2.2 | 36 | 2837 | 0.104 | 0.717 | 0.212 |
| 3.3 | 38 | 2632 | 0.112 | 0.772 | 0.229 |
| 4.4 | 40 | 2444 | 0.122 | 0.841 | 0.248 |
| 5.6 | 42 | 2270 | 0.131 | 0.903 | 0.268 |
| 6.7 | 44 | 2111 | 0.142 | 0.978 | 0.289 |
| 7.8 | 46 | 1964 | 0.153 | 1.054 | 0.312 |
| 8.9 | 48 | 1828 | 0.165 | 1.137 | 0.336 |
| 10.0 | 50 | 1702 | 0.178 | 1.266 | 0.362 |
| 15.6 | 60 | 1206 | 0.256 | 1.764 | 0.522 |
| 21.1 | 70 | 867 | 0.363 | 2.501 | 0.739 |
| 26.7 | 80 | 633 | 0.507 | 3.493 | 1.032 |
| 32.2 | 90 | 468 | 0.698 | 4.809 | 1.422 |
| 37.8 | 100 | 350 | 0.950 | 6.546 | 1.933 |
| 43.3 | 110 | 265 | 1.275 | 8.785 | 2.597 |
| 48.9 | 120 | 203 | 1.693 | 11.665 | 3.448 |
| 54.4 | 130 | 157 | 2.224 | 15.323 | 4.527 |
| 60.0 | 140 | 123 | 2.890 | 19.912 | 5.881 |
| 65.6 | 150 | 97 | 3.719 | 25.624 | 7.573 |
| 71.1 | 160 | 77 | 4.742 | 32.672 | 9.656 |
| 76.7 | 170 | 62 | 5.994 | 41.299 | 12.203 |
| 82.2 | 180 | 50 | 7.512 | 51.758 | 15.295 |
| 87.8 | 190 | 41 | 9.340 | 64.353 | 19.017 |
| 93.3 | 200 | 34 | 11.526 | 79.414 | 23.468 |
| 98.9 | 210 | 28 | 14.123 | 97.307 | 28.754 |
| 100.0 | 212 | 27 | 14.696 | 101.255 | 29.921 |

**FIGURE 6-1.** This chart shows you how many cubic feet of moisture must be boiled away per pound of water.

## QUICK EVACUATION

BTU Buddy explains that there are two key elements for a quick evacuation, whether a large amount of vapor is present, or a small volume of vapor needs to be removed in a hurry: large-bore leak-free tubing, and the correct size and type of vacuum pump.

When a large vacuum pump is used for small gauge lines with possible leaks; such as 1/4-inch typical gauge line hoses with crimped fittings, O-ring seals, and Schrader valve connectors; you can expect it to be slow with the possibility of leaks.

"Bob, you have a special gauge manifold that can be used with your vacuum pump for quick evacuation," BTU Buddy says. "This manifold would speed up the evacuation, and time is money saved for the customer. This gauge manifold is the four-line manifold that includes both a refrigerant cylinder connection and a vacuum pump connection, as shown in Figure 6-2. The vacuum line is 3/8-inch instead of the typical 1/4-inch.

"Most manifold gauges have a small bore where the valve operates, as seen in Figure 6-3. The four-line manifold has diaphragm valves that have a larger bore, as shown in Figure 6-4. Many technicians use

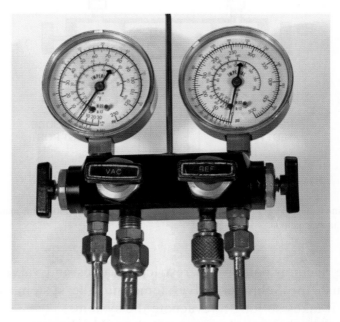

**FIGURE 6-2.** This gauge manifold has four valves and lines. Notice the refrigerant cylinder and the vacuum pump have their own line and valve and you can switch from the vacuum pump to the refrigerant cylinder without disconnecting the lines.

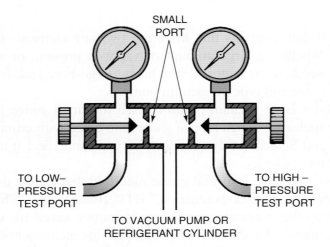

**FIGURE 6-3.** This gauge manifold has very small valve ports which will restrict evacuation.

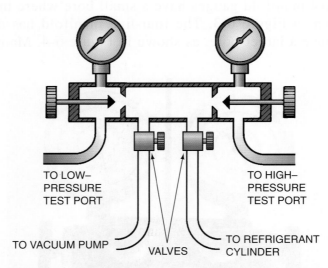

**FIGURE 6-4.** This four-line manifold has diaphragm valves with extra-large gauge ports. This manifold also has a 3/8-inch vacuum pump line to aid in evacuation.

large vacuum pumps and restricted-valves and valve stem depressors to push the Schrader valve stems into the system, and they wonder why it takes so long to evacuate the system.

"To speed evacuation, remove the valve depressors at the end of the gauge lines," says BTU Buddy. "Remove the valve stems from the Schrader valves on the liquid and suction line. Evacuate the system and when the evacuation is complete, pressure the system back to 0 psig with refrigerant and replace the valve stems. Place a valve stem depressor valve, Figure 6-5, at the end of each gauge line, reconnect the gauge manifold to the system, and charge it. This will give a quick evacuation to a very deep vacuum."

The four-line manifold also prevents suction of air into the manifold while disconnecting from the vacuum pump to connect to the refrigerant cylinder, as depicted in Figure 6-6. "There is no way to disconnect a typical manifold from a vacuum pump without pulling air into the valve section," remarks BTU Buddy. "Once that air is in there, **it can't be purged from the manifold while connected to the system that is under a vacuum.** This air in the system equals only a very small amount, but the manufacturer calls for 'no air,' not a little bit."

"You're right," says Bob, "I have that manifold in the truck but I've never used it because I wasn't sure how to up until now!"

BTU Buddy says, "Well, get the manifold and vacuum pump and let's use it to its maximum benefit."

Bob gets the tools, and solders another 1/4-inch fitting into the gas line on each heat pump line set, and then starts the vacuum pump on the upstairs unit. As mentioned above, Bob had removed the valve stem depressors from the gauge manifold lines so the lines had their full 1/4-inch bore without any restrictions.

BTU Buddy suggests, "Let's leave the line set capped at atmospheric pressure when we have finished dehydrating it. The construction crew can then finish their job just as it should have been, and you can move on to more service calls."

Bob says, "That sounds good to me."

The vacuum pump runs for about 15 minutes and the micron gauge goes down to 5,000 microns (the same as 5 millimeters). BTU Buddy then suggests that Bob shut off the vacuum pump and look for a rise in pressure. The pressure goes up a little and stops. BTU Buddy then suggests that Bob allow

**FIGURE 6-5.** This valve stem depressor can be used to depress the valve stem or actually back it out into the valve body during evacuation.

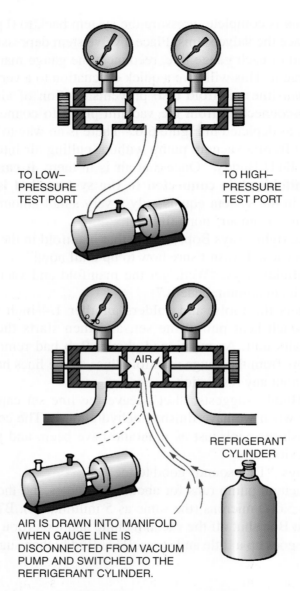

TO LOW–
PRESSURE
TEST PORT

TO HIGH–
PRESSURE
TEST PORT

AIR

REFRIGERANT
CYLINDER

AIR IS DRAWN INTO MANIFOLD
WHEN GAUGE LINE IS
DISCONNECTED FROM VACUUM
PUMP AND SWITCHED TO THE
REFRIGERANT CYLINDER.

**FIGURE 6-6.** This three-line two-valve manifold will allow air to be trapped in the center port of the valve when the technician changes from the vacuum pump to the refrigerant cylinder. Once this air is pulled into the gauge manifold, it will go into the system.

some nitrogen into the lines to help absorb any moisture in the system. Bob allows nitrogen into the system until the pressure is about 10 inches of vacuum on the manifold vacuum gauge, and then he starts the pump again. The system quickly pulls down to 5,000 microns and keeps on going down.

"We have this one system evacuated," says BTU Buddy. They let it pull down to 250 microns and valved off the vacuum pump. After about 10 minutes the vacuum gauge rises to about 350 microns and remains steady. Bob then pressures the piping up to about 10 psig and puts the caps on the two connectors that he had used for the vacuum.

Bob then connects the vacuum pump to the second system, the one that seems to have more water in it. The vacuum pump runs for about 30 minutes and only pulls down to 29 inches on the compound gauge and seems to stall. The micron gauge was not even reading because its scale starts at about 5,000 microns. The pump continues to run. BTU Buddy comments, "The vacuum pump is working; you can hear the tone change. Valve it off and listen to what it does."

Bob shuts the valve to the vacuum pump and it begins to run more quietly. "The pump is actually pulling water vapor from the system," BTU Buddy explains. "It's going to take time."

About an hour later, Bob notices that the oil level is rising in the vacuum pump. It's also turning white or milky colored. "What's going on?" Bob asks BTU Buddy.

"This is normal; you are pulling water from the system. This water is displacing the oil in the vacuum pump. It's time to change it," BTU Buddy replies.

Bob valves the vacuum pump off and drains the oil. He adds new oil and starts the pump again.

BTU Buddy says, "You must take very good care of the vacuum pump. If you don't change the oil periodically under these circumstances, the water will completely displace the oil and ruin the pump. You'll have to change the oil again. Actually, when the system begins to pull down, you'll have to change it two or three times to reach the desired vacuum. This pump has a gas ballast that is supposed to prevent moisture from entering the pump crankcase, but there's too much water in this system to stop all of it. After you're through with this system, you'll need to change the oil several times to clean the pump up for the next job. Keep saving the old oil and we'll dispose of it in the proper way. You're lucky this pump only takes a few ounces of oil, unlike the pumps of the past which would require you to collect a gallon or more of used oil from a job like this."

About an hour later, the system pressure begins to fall and register on the micron gauge. The oil in the vacuum pump was milky again and rising in the crankcase. BTU Buddy advised, "It's time to change the oil until the vacuum pump is clean."

Bob changes the oil until the vacuum pump pulls the vacuum gauge down to 100 microns, with the valve between the micron gauge and the system. BTU Buddy says, "The vacuum pump is clean and ready to go for the deep vacuum."

Bob connects the pump and starts it. The pressure keeps falling until the system is down to 300 microns. Bob adds nitrogen to the system until the pressure is about 10 psig, he removes the gauges, and caps the lines for the construction crew.

BTU Buddy advises, "Change the oil in the vacuum pump again while it's hot, as the residual oil and any moisture will drain from it completely."

Bob changes the oil and puts the vacuum pump back in the truck compartment. As they leave the job, BTU Buddy says, "You did a good job with this system. It's as dry as when the manufacturer sent it out, and it will provide long service life. Your company has provided you with the best tools that allow you to do a professional job."

# Changing a Badly Burned Compressor

This service call begins with a call from the dispatcher about a small business with a 5-ton single-phase heat pump that is not cooling. It is 90°F outside and the employees are really getting hot.

Bob arrives at the job site and goes in to talk to the manager. The manager tells Bob that the system was working yesterday afternoon, but when she turned the unit on this morning, it never cooled down.

## A PRELIMINARY CHECK

Bob goes to the thermostat, which is set to 73°F, but the temperature indicator shows 80°F. He can hear the indoor fan running, so he goes to the closet where it's located. All seems to be normal except that the suction line is not cold.

He now decides to go on the roof, where the outdoor unit is located. He leans a ladder against the building and ties it to the gutter so it will remain steady and not blow down. He then gets his tools and a voltmeter and goes up on the roof. The first thing he discovers is that the outdoor unit is not running.

He removes the cover to the control panel and can hear the compressor contactor humming, so he knows the thermostat is calling for the compressor to run. This puzzles him for a minute until he remembers that the power supply for the low voltage system is in the indoor unit, so this power supply is independent from the outdoor unit. Bob uses his voltmeter and checks the voltage at the line side of the contactor. There is no voltage. He then raises the cover to the breaker panel and checks voltage on the line side of the breaker. There is 234 V showing. He then checks the load side of the breaker and there is no voltage.

Bob starts to reset the breaker and BTU Buddy appears, "Do you think it's a good idea to reset the breaker without a further check?"

"It's probably just a random breaker trip," Bob reasons.

BTU Buddy then says, "It's easy to switch to ohms and check out the circuits."

Bob switches the meter to the ohms scale and checks from the load side of the breaker to ground, and there is a dead-short. Bob remarks, "That was a good idea. If I had reset the breaker, it may have done more damage."

BTU Buddy notes, "Now you must figure out what the ground circuit is. It will most likely be the fan or the compressor, as it's unlikely that a wire has gone to ground."

## LOOKING FOR THE GROUNDED CIRCUIT

Bob goes to the indoor thermostat and turns it to the off position. This takes the low voltage off of the outdoor compressor contactor. The reason that he could check ohms at the load side of the breaker was that the contactor's low-voltage coil was energized, making a circuit to the compressor. Bob now places one lead of his meter on a copper line in the condensing unit and the other on the load side of the contactor, and 100 ohms resistance registers on the RX1 ohms scale. He then disconnects the compressor leads and touches one of the meter leads to a compressor terminal. There is 100 ohms resistance. The compressor is grounded, as seen in Figure 7-1.

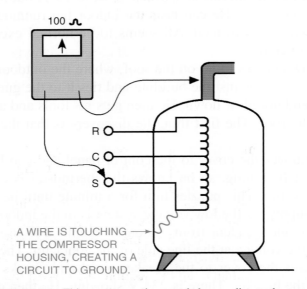

**FIGURE 7-1.** This compressor is grounded, according to the meter.

Bob calls the office for an assistant and then starts for the truck to get the needed equipment from the roof. BTU Buddy says, "Now is a good time to check the crankcase heater. It could be the cause of the motor burn."

Bob asks, "How can that be? A motor burn is an electrical problem and a crankcase heater can only cause a mechanical problem."

"Well," says BTU Buddy, "if the crankcase heater fails, you're right. It will cause a mechanical problem at startup, such as ruined valves, thrown rods, and such, due to liquid slugging. Without the crankcase heater, liquid refrigerant will migrate to the compressor crankcase. Most of the refrigerant charge can migrate to the crankcase overnight. Can you visualize a rod being thrown and some of the parts hitting the motor windings and causing a motor burn?"

"Well, yes," says Bob. He reaches down and touches the compressor housing and remarks, "The compressor is hot."

BTU Buddy suggests, "Tape the compressor terminal leads and turn the power on, and see if the crankcase heater is getting hot."

Bob does this and feels the crankcase heater at the base of the compressor and says, "It's not getting hot. The heat in the compressor must have been from the motor ground." BTU Buddy says, "The motor may have burned and then grounded. This would create a hot motor and then a ground."

Bob says, "I don't know what to think now, I'm confused."

BTU Buddy says, "Suppose the compressor started up with a thrown rod that hit the windings and caused the motor to start to cook. There may not be enough amperage to trip a breaker or an overload, but as it cooked, a circuit to ground was created and when that happened, it tripped the breaker. Can you visualize that?"

Bob says, "You really have to think to put the pieces together to clarify that. We're probably classified as service detectives."

## CHANGING THE COMPRESSOR

Bob and his assistant Andy go to the truck and bring up gauges and a recovery machine along with a recovery cylinder. As he connects the gauges to the system, he gets a smell of the refrigerant from the system. "Wow, that refrigerant really stinks," says Bob.

BTU Buddy says, "That's a sign that you're dealing with a very bad burn. When a system burns and cooks while it's running, the oil and refrigerant are saturated with the soot from the cooking. A running burn usually blows the soot down the discharge line to the condenser. A standing burn may throw the contaminants up the suction line toward the evaporator. This will require thorough cleanup procedures."

Bob recovers the refrigerant and makes a list of supplies needed, including:

1. Compressor
2. Refrigerant
3. Brazing material and flux
4. Suction line filter-drier (acid-removing type with pressure-drop connections), pipe, and elbows to pipe it in
5. A compressor contactor, start capacitor, and run capacitor in case they're needed.

BTU Buddy says, "How about the two liquid line driers, extra vacuum pump oil, and nitrogen?"

"You're right," says Bob, "this is a heat pump system and it will require two acid-removing filter-driers as shown in Figure 7-2. When I evacuate the system, the vacuum pump oil will get really dirty and we can use the nitrogen for making the braze connections."

BTU Buddy says, "Many technicians use a bi-flow drier to save time, as seen in Figure 7-3; that's your choice." He adds, "You can also purge all of the lines with nitrogen to blow any loose soot out from the lines. How will you know if any of this contamination reached into the suction line accumulator and four-way valve?"

Bob says, "I don't know, how?"

BTU Buddy replies, "If we discover a lot of soot in the suction line when you cut it loose, you should assume that soot moved into them also. So it would be a good idea to bring a new accumulator and four-way valve, as well."

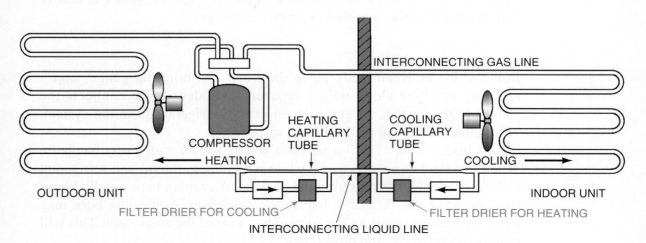

**FIGURE 7-2.** Note the liquid line drier that is piped parallel to each metering device.

COOLING          HEATING

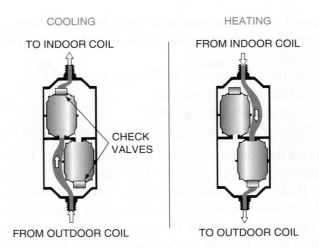

**FIGURE 7-3.** This is a bi-flow drier. Actually it is two driers in one shell with check valves to prevent backwash of a full drier.

"Boy, this is a lot to remember," says Bob.

Bob returns with all of the supplies that have been listed and talked about. The system is in a vacuum from the refrigerant recovery. BTU Buddy suggests that Bob bring the system pressure up to atmospheric pressure with nitrogen.

"Why?" asks Bob.

"The nitrogen keeps oxygen from entering the system and prevents oxidation to the piping and parts from the remaining acid that must be in the system," says BTU Buddy.

Bob connects the nitrogen and brings the system to atmospheric pressure.

BTU Buddy recommends, "Let's make a plan for this changeout." The plan is:

1. Get all parts to be used, or that may be used, laid out within easy reach.
2. Cut the compressor suction line loose as close to the compressor as you can, using tubing cutters, not a hacksaw. Examine the condition of the inside of the suction line.
3. Cut the discharge line loose as close to the compressor as you can, again using tubing cutters. Examine the inside of the discharge line.
4. There are gauge lines connected to the liquid line and the gas line outside the unit. Blow nitrogen through the liquid line and see what comes out.
5. Replace the compressor and the four-way valve and accumulator if needed.
6. Install both the liquid line driers.

7.  Install the suction line acid-removing filter-drier.
8.  Pressure test with a small amount of R-22 and nitrogen pressure up to 150 psig.
9.  Blow the nitrogen and R-22 from the system.
10. Evacuate to a deep vacuum.
11. Charge and start the system.
12. Check for any other signs that may have caused the motor burn.

Bob cuts the compressor loose at the suction line. There is a slight amount of nitrogen pressure in the system. "This nitrogen really smells bad," says Bob.

BTU Buddy says, "Don't breathe too much of that; it's toxic. Stay upwind."

Bob removes the burr from cutting the suction line loose and takes a white rag and wipes inside the pipe. It comes out oily but clean.

"Well, I don't believe you'll need the new four-way valve and accumulator," BTU Buddy comments.

"That's good news for the customer," says Bob.

Bob cuts the discharge line loose and can tell right away that it's full of soot. It's almost closed up.

BTU Buddy says, "That is a sign that the compressor burned while running. Most of the contamination will be confined to the high pressure side of the system. Now blow some nitrogen through the liquid-line gauge connection. Place that clean rag over the end of the discharge line to catch debris."

Bob follows his directions and a big glob of oil and soot deposits in the rag. He gives it several blasts of nitrogen until only clear oil droppings come out. Nitrogen is also coming out of the loose suction line. It's clear.

"I think you have the system as clean as you can get it. It's ready for the compressor and then the driers. If you let the nitrogen trickle through, it will keep the compressor and driers purged and prevent contaminating them with the humidity in the air," says BTU Buddy.

The old compressor is removed and the new one set in place. A short stub of pipe is used to connect the new compressor in order to avoid having to remove the old braze filler metal from the compressor stubs. The new short length of pipe and a coupling make a much cleaner job.

Bob then solders in the liquid-line driers, one for each metering device, Figure 7-2. Bob then solders the suction line filter-drier into the permanent suction line, between the four-way valve and the accumulator. The constant trickle of nitrogen prevents oxidation inside the system while brazing-in the high-temperature connections and keeping atmosphere from entering the system.

Bob then installs the crankcase heater. He checks the compressor contactor and finds the contacts in bad shape after the motor burn, so he changes it. The capacitors all seem to be in good shape so he runs a capacity test on them with his capacitor tester. They are well within range so he doesn't change them.

## EVACUATING THE SYSTEM

Bob evacuates the system using a four-line gauge manifold. He now has the system at atmospheric pressure with a holding charge of refrigerant in the system. He has turned the refrigerant cylinder to allow liquid into the liquid line and he opens the valve to the system. When the liquid stops flowing, he closes the gauge manifold valve to the high side, and starts the system. It is 92°F outside and it is hot inside the office.

The suction pressure is running at 156 psig (corresponding to 55°F evaporating temperature R-410A) and the discharge pressure is 506 psig (corresponding to 135°F condensing temperature). The pressures are about right for an overloaded system and a hot pull-down.

## ADJUSTING THE CHARGE

BTU Buddy asks, "How are you going to adjust the charge to make sure that it's correct?"

Bob says, "I'm going to clean up everything while the building temperature decreases to the normal range, and I'm going to charge the system to the proper superheat for a line set of 10 feet." (This procedure was discussed previously, in Chapter 3.)

The office is beginning to really cool down by the time Bob picks up all of his tools. Bob adjusts the charge by checking the superheat at the outdoor unit until it is about 10°F. He then checks the subcooling of the liquid line and it's 12°F. The pressures are about right for the conditions: the suction pressure is 90 psig (evaporating temperature of 42°F) and the discharge pressure is 332 psig (condensing at 125°F).

Bob momentarily reverses the heat pump to heat, to verify that the four-way valve will change over (by jumping from the "R" terminal to the "O" terminal at the outdoor unit). When the valve changes over to heat, he immediately removes the jumper. This system cannot be allowed to run in the heat mode at temperatures this high.

BTU Buddy compliments Bob, saying, "You have done a great job of getting this compressor changed in a very reasonable time frame. I like that you explained your procedures to the customer, that gives them confidence in you and your company. When you make the company money and keep the customer happy, you become a valuable employee. It's called craftsmanship. Become a good craftsman and you'll be a valuable asset wherever you go."

Bob says, "I wish every technician had a BTU Buddy."

BTU Buddy says, "They do. It's in the texts they studied, the instructors they had, the friends they have in the industry, and the pride they take in their workmanship. All technicians have a BTU Buddy at their disposal. They just have to use it."

# Burned Four-Way Valve Coil on a Heat Pump

The dispatcher contacts Bob with a service call from the manager of a small retail store. She is complaining that the store is too hot; the system seems to be heating instead of cooling. Bob advises the dispatcher to call the store and tell the manager to turn the system off until he can get there. It will be a couple of hours.

As Bob is driving over to the job about two hours later, he begins to think about the problem. He's not familiar with this client so he must check the job out completely to become familiar with it. He arrives to find a hot and agitated customer.

The store manager meets Bob at the truck and tells him that she needs some relief inside. It is a clothing store and customers don't like to look at or try on clothes in a hot, sweaty atmosphere; this is hurting business.

Bob realizes that he must maintain his professional manner and get the customer's work done in a timely fashion. There is as much invested in keeping the manager happy as in doing good work. BTU Buddy then appears and says to Bob, "Since you have an irritated manager, ask for her opinion and she may feel like part of the call and actually help you. This will help get her on your side. There is no need to struggle with an unfamiliar service call and an angry manager at the same time."

Bob goes into the store with the manager and asks her to review what happened. She says, "I arrived at work this morning at about 8:00 a.m. and turned on the system like I usually do. It was warm in the store when I arrived. A few minutes later, I was working at my desk and noticed that my office kept getting hotter and hotter. I reached up toward the ceiling where the cool air normally blows down and it was warm."

Bob says, "Show me what you did."

"Great work," BTU Buddy says to Bob, "she is on your team now. She'll show you the thermostat and tell you where the system components are. You didn't know any of this when you came to the job and, she's helping. Sometimes it's hard to listen to the customer, but oftentimes it's a big help."

The manager says, "The system is a heat pump and has worked great for the seven years that I've been here. All we do is change the filters every month. We haven't had a service call in the last seven years." She then says, "There are two pieces to the unit; one is in a closet and the other is on the roof. You can access the roof through the ladder in the storeroom. Just push the hood access cover up as you go up the ladder. I learned that from the plumber who had to do some work on the plumbing vents."

Bob says to the manager, "I'm going to the truck to get some tools and then we'll turn it on and see what it's doing."

## CHECKING OUT THE UNIT

When Bob comes back with his tool pouch and electrical meter, he turns the thermostat to "cool" and heads for the roof. When he gets to the roof, he notices that the unit is running. When he puts his hand in the air stream leaving the coil, he discovers that it's cool, not warm. He knows this means that the unit is running in the heating mode, not the cooling mode, but the big question is why?

He shuts the unit off using the disconnect switch because he knows that it's very hard on a heat pump to operate when the outdoor air is warm, and it's 85°F. The heat pump will be overworked when it is absorbing heat from an outdoor temperature of 85°. Soon it will trip off because of over-load, which is hard on the compressor. He will leave the disconnect switch off while investigating the problem.

Bob now starts the troubleshooting process. He knows that he must not run the unit unless absolutely necessary, but he must find the problem. BTU Buddy asks him, "What are the possibilities?"

Bob says, "The four-way valve may be stuck in the heating position. The thermostat or related wiring could be a problem, too."

BTU Buddy says, "Why don't you look at the wiring diagram and see whether the four-way valve is energized in cooling or heating?"

"What difference could that make?" Bob asks.

"Well," answers BTU Buddy, "the system should not have been switched to the heating mode since it hasn't been cool enough to cause a call for heat. So whatever happened had to happen in the cooling mode, unless the operator set the thermostat to call for heat. Think about the possibilities."

Bob removes the panel with the wiring diagram and reviews it. "The four-way valve is energized in the cooling mode," he says. See Figure 8-1 for a simplified wiring diagram of this unit.

"What would happen if the four-way valve coil were to malfunction, or the wiring calling for the valve to be energized were to cause no voltage to be applied to the valve coil?" asks BTU Buddy.

"Well," says Bob, "it looks to me like the unit would operate in the heating mode."

BTU Buddy explains, "It's easy to check the four-way valve coil to see if it has power supplied to it. If it doesn't, then check it for continuity."

Bob is able to check power to the four-way valve without turning on the power at the outdoor unit, because the indoor unit contains the power transformer and the four-way valve coil is 24 volts. He finds that there was power at the four-way valve, but the coil was not warm; it was not pulling current.

BTU Buddy says, "There is one other thing you can do at this point to see if the coil is functioning. If you hold a screwdriver blade up close to the coil, it will act like a magnet if it's working. When you notice that it's not warm, you know it isn't working, but if it had just turned on, it wouldn't be warm yet anyway, so the screwdriver method can be used."

Bob then removes the wires from the four-way valve coil. Since it's only 24 volts, he doesn't need to go downstairs to the air handler where the 24-volt transformer is, to turn the power off. He then checks it for continuity with an ohmmeter and finds that it has an open circuit.

Bob says, "This is the problem. The valve coil has an open circuit and has caused the unit to change over to heat."

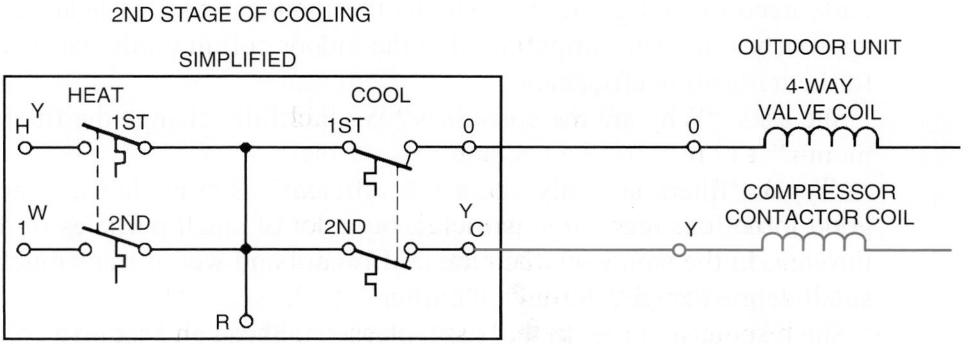

**FIGURE 8-1.** Notice that the thermostat controls the four-way valve from the "O" terminal, which is the first stage of heat. The interconnecting wire is also usually routed in orange.

Bob changes the coil and replaces the wires to the coil, and hears the valve's pilot valve click. This is a good sign that the unit will now function correctly. He starts the unit up and hears the four-way valve change over to cooling when the pressure inside the unit starts to build up. He then asks, "Why didn't the unit change over immediately?"

## HOW THE FOUR-WAY VALVE WORKS

BTU Buddy explains, "The four-way valve is actually two valves in one. The solenoid coil controls what is known as a pilot valve. The pilot valve directs the internal gases inside the main valve body piston to position the piston to either heat or cool. Figure 8-2 shows the internal workings of a four-way valve. When discharge pressure is routed to one end of the piston, low pressure will be on the other end and the valve will shift positions. For example, suppose high pressure is on the right-hand end. The piston will move to the left and discharge gas will be directed to the indoor coil, and heating will occur. This is what happened in this system. When the coil is de-energized, discharge gas is routed to the right-hand end of the valve. When the coil failed, it was the same as the thermostat calling for heat."

BTU Buddy adds, "Let's make sure the unit is cooling, by feeling the large line going downstairs. You can always tell which mode the heat pump is in by touching the large line. When it's cool, it is in cooling mode; when it's hot, it is in heating mode. You can also feel the heat in the air leaving the outdoor coil. That heat is coming out of the store."

Bob touches the large line and it's cold. Then he goes downstairs to report to the manager. When he tells her what he has done, she says, "Thanks for the prompt and efficient service call."

Bob has oiled the motors and checked the coils, and finds that the coils need cleaning. He explains to the manager that as heating season approaches, it's very important that the indoor coil in particular be cleaned for better heating efficiency.

She asks, "Why are the coils dirty? We faithfully change the filters every month."

"Typical filters are only about 5% efficient," Bob explains. "They do a good job of catching large particles, but a lot of small particles of dust get through. In the summer, when the coil is cool and wet, it traps much of the small debris that gets through the filter."

She responds, "I see. In that case, please go ahead and set up a coil cleaning for either early morning or late afternoon, so we can avoid any interruption in business hours."

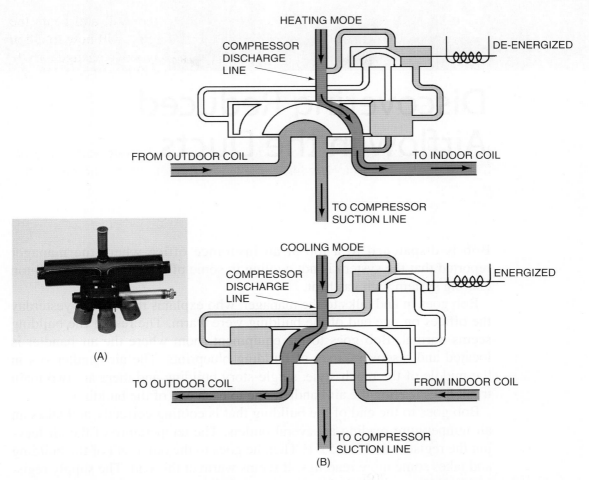

**FIGURE 8-2.** This photo is of a four-way valve with no coil. The top cutaway is in the heating mode and the bottom cutaway is in the cooling mode. Notice the small piping from the pilot solenoid valve which directs the hot gas to the right end of the main valve piston for heating and the left end for cooling.

Bob sets it up for the following day at closing time, when he will have a minimum of overtime and won't disrupt the business. The manager thanks Bob for his service and professionalism.

While driving from the job, BTU Buddy says, "You really did a nice job for that client. Plus, you booked extra work by looking around. The manager appreciates the fact that you were looking after the equipment, and your company appreciates you bringing in more business. This makes you a valuable asset to both your customer and your service company."

# Discovering Reduced Airflow in the Ducts

Bob is dispatched to a job at an insurance office where the manager reported that the system is not cooling in some of the offices. It is a 20-ton cooling system with gas heat.

Bob goes in and talks to the manager, who explains that all day yesterday the offices on one end of the building were warm. The rest of the building seems normal. Bob goes to the equipment room where the air handler is located and discovers a set of building blueprints. The air handler sits in the middle of the ranch-style, single-story building and there are two main trunk lines leaving the air handler, one to each end of the building.

Bob goes to the end of the building that is cooling correctly and takes an air temperature reading at several outlets. The temperature of the air leaving the registers is 55 to 57°F. Then he goes to the other end of the building and takes some more readings. It seems warm at this end. The supply registers read the same as the ones at the cool end of the building, but the space temperature is 78°F. It is warm, for sure.

Bob gets out his air velocity measuring instrument, a velometer, Figure 9-1. He takes some velocity readings in the cool end of the building and discovers them to be averaging about 250 feet per minute (fpm). He then goes to the warm end of the building and finds the velocities to average about 100 fpm. There is definitely a reduced airflow problem, and he must determine what it could be.

BTU Buddy then arrives and asks, "What do you think the problem may be?"

"I don't know," Bob says.

BTU Buddy asks, "Are there any dampers in the air system that could be closed?"

Bob looks and says, "Yes, there are two dampers shown on the print, one on each main run of duct. They should be located overhead, here in the equipment room."

(A)

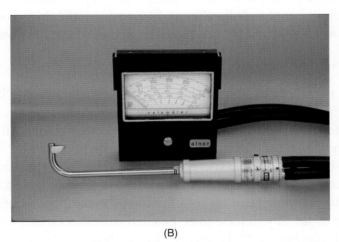

(B)

**FIGURE 9-1.** Two types of velometers for checking air velocity.

Bob locates the dampers, but the damper handles show full flow in each direction to each run of duct, and the handles do not look as though they've been disturbed.

BTU Buddy suggests to Bob that he check the fan amperage. "If the fan motor is not pulling enough amperage, that could be an indication of reduced airflow."

Bob checks the amperage. It's a 3-phase motor and the amperage is running 8 amps per phase. The full load rating on the motor is 12 amps.

BTU Buddy says, "The fan's not running under a load so that's another indication of reduced airflow. See on the motor disconnect switch where someone has marked 12 amps as the recorded running amps? It's a good idea to make those notes on the equipment so you have a reference point any time there is a question.

"You have narrowed the problem down to an airflow problem," he continues. "Now, we know that it must be in the supply duct because it seems to only affect one end of the building. What do you think you should do next?"

Bob scratches his head and says, "I can't imagine what it could be."

BTU Buddy asks, "How is the duct insulated?"

Bob responds, "With duct lining, on the inside. Could it have come loose and dropped down, and if so, where?"

"Well," comments BTU Buddy, "I don't see any weld spots on the outside of the duct where the duct liner tabs have been spot welded. It could be that the duct liner was fastened up with glue only. For some reason, the glue may have come loose." BTU Buddy then says, "Start checking the air registers closest to the air handler and move down the duct. You may discover an airflow change and be able to better predict where the flow is reduced."

## TRACKING DOWN THE AIRFLOW PROBLEM

Bob measures the air at the first two registers and it's averaging about 250 fpm. When he moves to the number three register, the airflow drops to about 100 fpm. He exclaims, "I think we've found the section of duct where the restriction is, but what can we do about it?"

BTU Buddy says, "There's only one thing you can do; you must cut an inspection hole in the duct and take a look. Cut the hole to a size that allows you to get your head up into the duct—make it good and square, with no sharp edges. It will have to be patched with sheet metal when you finish."

Bob gets his safety glasses and electric shears ready. He then lays out a square where he'll cut the hole. Meanwhile, he calls the office and asks them to send out some sheet metal panels of the correct gauge. He asks that the sheet metal foreman cut the panels and make a cross brake in them to give them strength and to punch holes around the perimeter for the screws to fasten the panels. He also asks for a gallon of mastic duct sealer to make the panels airtight.

BTU Buddy asks, "How are you going to fasten the duct liner back?"

Bob then tells the office to send some contact cement for fastening duct liner, along with some sheet metal tabs that can be stuck to the inside of the duct for the duct liner. Figure 9-2 shows a sheet metal tab that sticks on the duct. The liner is pushed over the pin and a tab is then pushed down over the pin.

Bob shuts the system down to stop all airflow while he's cutting. He then cuts out the inspection hole and looks in the duct. About 3 feet downstream he finds the problem. Sure enough, the liner has come loose and about 2 feet of it has rolled up, blocking most of the airflow.

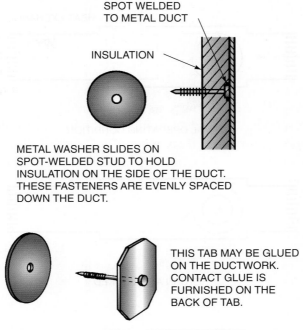

SPOT WELDED
TO METAL DUCT

INSULATION

METAL WASHER SLIDES ON
SPOT-WELDED STUD TO HOLD
INSULATION ON THE SIDE OF THE DUCT.
THESE FASTENERS ARE EVENLY SPACED
DOWN THE DUCT.

THIS TAB MAY BE GLUED
ON THE DUCTWORK.
CONTACT GLUE IS
FURNISHED ON THE
BACK OF TAB.

**FIGURE 9-2.** These two kinds of sheet metal tabs may be used to fasten duct liner. One is spot welded and the other is glued on. Both use a slip-on tab to hold the liner.

Bob then cuts another inspection hole where the liner is loose. He finds that the liner is in fact only glued. He then applies glue to the liner and the duct and lets it set for just a few minutes to allow the glue to become tacky. He has mounted several tabs in the same area to help hold the liner down. When the glue is ready to be contacted together, he rolls the liner back into place. He then puts the tabs on the pins to better secure the liner.

Bob then runs some self-tapping metal screws into the holes in the panel covers to secure them. Once he has installed the two panel covers, he starts the system and measures the airflow at the registers in the warm side of the building, and discovers an average of about 200 fpm. He goes to the cool side of the building and finds that the velocities average about 200 fpm there as well. All of the air seems to be going to the right places now. A check of the fan amperage shows 12 amps.

Bob then liberally applies the mastic sealer to the edges of the newly installed panels. The system will be airtight where he worked on it.

"I thought that a restriction in a supply duct would cause high fan amperage," he says.

BTU Buddy explains, "This is a common squirrel cage fan, and the current draw is in proportion to the quantity of air that passes through the fan.

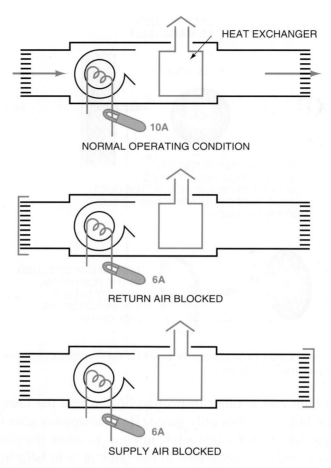

**FIGURE 9-3.** Notice that it does not matter how much you restrict the airflow through a basic squirrel cage fan; the current flow will be reduced.

The difference in the inlet and outlet pressure doesn't matter (Figure 9-3 shows a fan under different types of restrictions). It doesn't matter how the air is restricted, a reduced airflow will result in lower motor amperage. Because of this, the technician can measure airflow quantity using an ammeter. If the fan amperage is recorded at the time when maximum air is flowing, it can be used in the future for fan analysis."

Bob goes to the building manager and tells him what he found. Bob also schedules a return visit to clean the evaporator and condenser coils when mild weather comes.

BTU Buddy says, "You did it again; you brought more business to your company and did a good job on the service call. You are really becoming a professional."

# Fixing a Chronic Leaking Residential A/C Unit

The dispatcher calls Bob to do a "spring checkup" for a new customer. The dispatcher remarks that the customer wants his annual refrigerant fill-up.

Bob is on his way to the job, thinking about the call when BTU Buddy makes an appearance. He asks Bob, "What do you think about the customer calling it a spring checkup and annual refrigerant fill-up?"

Bob says, "It sounds like there's a leak that has never been fixed, and his previous service company just charged it every year without ever trying to find the leak."

BTU Buddy says, "That sounds about right. Some customers think that their A/C unit consumes refrigerant because the service company has been charging it for so many years. What do you think about approaching the customer about a repair rather than a refill?"

Bob agrees.

When they get to the job, Bob talks to the customer about the past service. The customer explains that the prior service company kept saying that it was cheaper to charge the unit rather than to find and repair the leak. He says they only added about 3 or 4 pounds of refrigerant per year.

Bob explains, "That's true for this year; it would be cheaper for me to add refrigerant than to find the leak and repair it, except that someone will have to add refrigerant again next year and the year after that. It's not good for the unit to run with a low refrigerant charge, because the refrigerant is what cools the compressor. The unit is less efficient and uses more power per BTU output with a low charge of refrigerant.

"If it were my own unit, I would repair it, in order to have a more efficient unit that will last longer with a correct refrigerant charge that stays in the unit. I wouldn't have to keep adding refrigerant every year, and I would also be protecting the environment. Service technicians are allowed to keep charging a system of residential size, but when it's done to hundreds of systems each year, a lot of refrigerant is unnecessarily vented."

The customer says, "It makes sense to go ahead and fix it. Find the leak and repair it."

## FINDING AND FIXING THE LEAK

Bob says to BTU Buddy, "You got me into this. What should we do first?" BTU Buddy responds, "Let's think about this and itemize the possibilities:

1. Run the unit and make sure it really has a low charge.
2. If the charge is low, let's look for the leak by shutting the unit down and testing the obvious places. Shutting the unit down will register the maximum pressure on the low side components. You cannot use an electronic leak detector while the unit is running because of the draft created by the fans. Service technicians leave all sorts of leaks around the service ports. We'll do that part of the leak test before applying gauges. If we apply the gauges first, we may cover up the service port leaks.
3. If we don't find the leak, we'll remove the charge and pressure the entire system up with trace refrigerant and nitrogen to the working pressure of the weakest link in the system. This is usually the compressor housing with a working pressure for R-22 of 150 psig. If the compressor can be valved out of the circuit, we can raise the pressure of the system up to 250 psig without concern.
4. If we still don't find the leak, we'll divide the system into two components and pressure each half for an overnight standing pressure test. This can be done by cutting the tubing, isolating the indoor system from the outdoor system, and installing test ports in the indoor section. The test ports on the outdoor unit can be used for testing that section. Let it stand overnight and see which one leaks down. This is a last-ditch sort of test; usually the leak can be found before this point. But the advantage of this test is that you can apply more pressure to the low-pressure side, which will amplify the leak."

Bob then asks, "Why is the compressor the weak link in the system? It pumps high-pressure refrigerant."

BTU Buddy explains, "The compressor is suspended inside the shell with the discharge line piped to the outside of the shell with reciprocating compressors." This is shown in Figure 10-1. He continues, "To really be safe, you should check with the compressor manufacturer for the working pressure of the shell. Newer refrigerants and scroll or rotary compressors will have different working pressures."

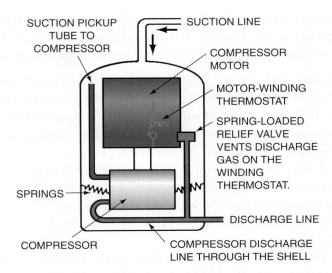

**FIGURE 10-1.** The internal piping of a compressor.

Bob starts the unit without the gauges so the gauge ports will not be disturbed. He goes to the evaporator section in the garage and notices there is ice on the short length of pipe between the expansion device and the coil. He says, "There is refrigerant in the system, but the charge must be low because of the ice."

BTU Buddy tells Bob, "Let it run for a few minutes to see if the ice clears up. Block some of the condenser airflow and see if the ice clears up. It's only 70°F outside; low head pressure may cause most of the refrigerant to be in the condenser. Block the airflow until you can feel warm air coming out the top of the condenser. You will then be assured of enough head pressure to move the liquid out of the condenser. When a unit has just been started for the first time, liquid refrigerant can be lying in the suction accumulator or the condenser. You must allow a little running time for it to be redistributed to the proper places in the system."

Bob gives the unit 15 minutes of running time while changing the filter and looking over the rest of the job. Then he says, "The system definitely has a low charge. There's still ice on the line before the coil."

BTU Buddy tells Bob to shut the unit down and says, "We'll let the pressures equalize and then leak-check."

Bob gets his electronic leak detector and says, "I don't seem to be able to use this detector with confidence. Would you give me some pointers?"

BTU Buddy explains, "The electronic leak detector's best quality makes it difficult for some people to use: it is super-sensitive. Most detectors will detect a leak rate of 1/4 ounce of refrigerant per year. If there is any refrigerant in the vicinity, it will sound off. Then you have to find where it came from. Here are some tips to improve your leak detecting abilities:

- Stop any drafts of air. You can use a piece of cardboard when outside to prevent wind from blowing the refrigerant away from the leak.
- Start high and move down. Refrigerant is heavier than air and will move downward. If you start low and get under a leak, you can't tell where it came from.
- Move very slowly, about 1 inch per second. I have seen technicians leak-detect an entire unit in 5 minutes and declare there was no leak. They may be moving past the leak and not knowing.
- Move over each field connection first, as that is the most likely place to find a leak.
- Pay attention for oil spots or dust spots. There is often a thin film of oil near a long-time leak and it will collect dust over time. There is almost always oil around the service valve connections from where the gauge hoses are connected.
- Make sure you do not let water get into the leak detector nozzle. When you get to the evaporator section, you can disconnect the condensate line and hold the detector probe up high, out of the water stream, and let the probe stay there for 2 or 3 minutes. If there is refrigerant detected, then remove the coil panels so you can check the piping and the coil tube turns.
- When you find the leak area, if it needs to be located exactly, use soap bubbles," Figure 10-2. "There are special soaps used for leak-detecting that are very elastic and blow big bubbles. They won't dry out like house-hold detergents. *Remember:* After using soap on piping, wash it off with water and a rag. If you don't, oxidation will begin at that point. If the pipe is copper, there will soon be a large green stain. Professionals do the job correctly and don't engage in bad habits."

Bob uses the electronic leak detector and goes over the gauge ports and the field connections on the outdoor unit. He then places the leak detector probe under the insulation for the suction line and finds no leaks. There are no connections between the outdoor unit and the indoor unit, so he moves on to the indoor unit. He first checks the field connections and finds no leak. He places the probe for the electronic leak detector in the condensate drain line in such a way that no water can enter the probe. He holds it there for about a minute when the probe picks up a leak.

Bob removes the panels for the coil and starts the search. First he has to wipe off the moisture from operating the coil, then as he checks the suction

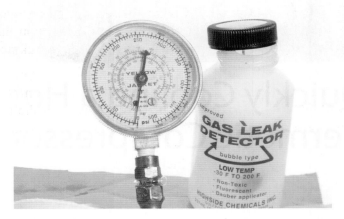

**FIGURE 10-2.** Using professional quality soap bubbles to find a leak.

line leaving the coil, he detects the leak. He applies a little soap and finds the leak at a faulty factory connection that can be easily repaired.

Bob then recovers the refrigerant and repairs the leak. He tells BTU Buddy, "That wasn't bad. Now the customer can have confidence that the unit will provide good service."

BTU Buddy asks, "How do you know that's the only leak?"

Bob says, "I found a leak and repaired it. Is there more that should be done?"

BTU Buddy explains, "You have probably found the only leak, but to be sure, let's evacuate the unit to about 25 inches of vacuum and pressure the system to 150 psig and let it stand overnight. When we come back tomorrow and the pressure is still there, you can be certain. We need to remove the refrigerant and use only nitrogen, to be sure that the temperature is not affecting the pressure."

Bob completes the process, leaving a set of gauges on the system. He then informs the customer of what he's doing. The customer understands and gives approval.

Bob was unable to come back the next day, but the following day when he arrives at the job, he heads straight for the gauges and finds them reading 150 psig. He is pleased to find that the pressure hadn't dropped.

Bob evacuates, charges and starts the unit, then explains to the owner what he has done.

The owner says, "I'm so glad you fixed it. That gives me a lot of confidence in you and your company."

As Bob is driving away from the job, BTU Buddy appears again and says, "It really makes you feel good when a customer says that to you. You did another professional job."

# Quickly Cooling a Hot Hermetic Compressor

When Bob receives his next call, the dispatcher tells him that a longtime customer is without air conditioning. It's a restaurant and the lunch hour is coming up in 3 hours. The owner wants it repaired as soon as possible.

Bob arrives at the job and goes onto the roof for a look at the rooftop unit for the large dining room. The condenser fan is running, but the compressor is not.

Bob takes the compressor compartment door off and looks inside. He tries to touch the compressor and quickly pulls his hand back. It is really hot. He notices that the compressor contactor is humming, so the control circuit is calling for cooling.

He then checks the voltage at the outlet to the compressor contactor and finds that it has line voltage from L1 to L2; it's a single-phase compressor. There's power to run the compressor, but it isn't running. Bob steps back and scratches his head.

BTU Buddy appears and asks Bob, "What do you think is wrong?"

Bob says, "It looks like the compressor is burned up. It's too hot to touch."

BTU Buddy asks, "What are you going to do now?"

Bob says, "Take down the model and serial numbers and order a new compressor."

BTU Buddy then asks, "What if it's just hot? If you change it out, when you get back to the shop the compressor will have cooled and will then run."

Bob asks, "What do you suggest?"

BTU Buddy explains, "Suppose you give the compressor an electrical check. It will only take a few minutes and it may tell you something. If the compressor is burned, it will likely have a circuit to ground. If it has an open winding, it may be the internal overload that has it off. If that's the case, you can cool the compressor and start it back up. The high temperature of the compressor may be a result of operating with a low charge. Remember, the suction gas cools the compressor; reduced suction gas results in a hot compressor."

## MAKING THE ELECTRICAL CHECKS

To check the compressor, Bob does the following:

1.  He turns off the power and locks the panel. He then checks the compressor from the terminals to ground and there is no circuit. The compressor is not grounded.
2.  He then checks from common to run. The circuit is open.
3.  He checks from common to start. That circuit is open.

Bob says, "This compressor has open circuits. It must be defective."

BTU Buddy notes, "You haven't done the next and most important test, from run to start."

"Why is that so important?" asks Bob.

BTU Buddy says, "Suppose the compressor is hot because of some other reason; for instance, an internal overload that may be open. If so, you would get the readings that you've already noticed; then from run to start there would be a circuit through those windings." Figure 11-1 shows a hot compressor with an open internal overload.

BTU Buddy asks Bob, "So now that we have discovered a hot compressor, what are our options to get the system operating by lunch hour for your customer?"

Bob says, "We could come back late this afternoon, after the compressor has cooled down, but I don't think they would be happy without cooling for a busy lunch hour. What do you suggest?"

BTU Buddy says, "Why don't you do a quick cooldown with water? That would cool it enough to start in about 30 minutes. While it's cooling, you can put gauges on the unit and see if you can find the problem."

**CAUTION** Make sure the water does NOT come into contact with any of the electrical components. Use some dry timber to stand on while completing the rest of the work with power on the unit.

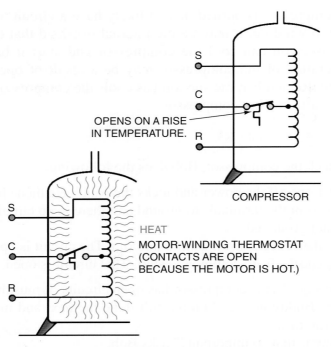

**FIGURE 11-1.** This compressor is hot and the internal overload is open. This allows an ohm reading through the run to start circuit only. There is no reading from common to run or common to start.

Bob gets a water hose and sets it to a slow trickle of water on the top of the compressor shell. Figure 11-2 shows where to direct the water flow. He then fastens gauges on the unit and discovers that there is no refrigerant in the system. Bob says, "This is why the compressor is hot. There's no refrigerant in this system. Why didn't the low pressure control shut the system off?"

BTU Buddy says, "Look at the diagram to see why."

Bob looks at the diagram and says, "This system doesn't *have* a low pressure control. How is it protected from a low charge such as this?"

BTU Buddy explains, "Many systems use the winding thermostat for low-charge protection. The motor will just warm up and shut the system off. It won't get hot enough to harm the compressor. When the winding thermostat shuts the compressor off, the large mass of the compressor and motor will keep it off for a long while. Now let's see if we can find out why it lost the refrigerant. The loss of charge had to happen overnight because the owner said the system was working last night. It's probably something obvious. First do a visual test."

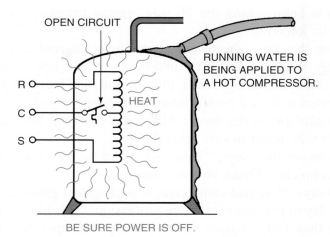

OPEN CIRCUIT

RUNNING WATER IS
BEING APPLIED TO
A HOT COMPRESSOR.

R

C

S

HEAT

BE SURE POWER IS OFF.

**FIGURE 11-2.** This compressor is being carefully cooled with water. CAUTION: Do not allow water to reach any electrical components. Use dry wood to stand on when starting the unit to keep your feet out of the water.

Bob looks at the system and finds a fresh oil slick where the discharge line is touching a piece of flexible electrical conduit. There is just a small pinhole size leak. He gets his torch and repairs the leak.

BTU Buddy comments, "The system never went into a vacuum, so evacuation isn't necessary. You can just charge the unit and move on. First, allow liquid refrigerant to enter the system through the liquid line until tank pressure is reached; this will be close to the correct charge. Then you should get set up to charge liquid into the low-pressure side of the system because the compressor is still going to be plenty warm. By using liquid metered into the low side of the system, you can further cool the compressor as you are charging it. This will reduce the chance of the compressor shutting off again because of high temperature."

"How do we know if the compressor is cool enough to start?" asks Bob.

BTU Buddy says, "Once the system is set up with the partial charge of refrigerant, get ready to charge the liquid into the low-pressure side and turn it on. If it starts, just finish the charge into the low-pressure side."

Bob gets everything ready. He finds some old forklift pallets behind the building and places two of them next to the unit because the roof is still wet from the water.

Bob turns on the unit disconnect switch and the unit, and the compressor starts. The suction pressure is still low, so he slowly allows liquid refrigerant to enter the suction line until the suction pressure reads about 65 psig, which is approaching the 70 psig normal pressure for R-22. The unit is doing its job now.

About that time, the owner comes onto the roof and says, "The unit is cooling. What a relief. You just don't realize how many customers will come into a hot dining room, turn around, and leave to go down the street. They want to have a comfortable lunch hour."

Bob explains to the owner what happened, and how he cooled the compressor with water to save several hours.

The owner then says, "That's what I'm happy to pay for; a thinking service technician. Thank you very much."

Bob says, "I've had some great schooling and guidance, and I'm working every day to become even more professional. Thanks for the compliment."

Bob then fully charges the unit using the super heat method, since the unit has a capillary tube metering device. He then replaces all of the panels and returns his tools to the truck.

BTU Buddy says, "You did another good job there. You built excellent customer relations. I have never understood why some service technicians don't take care of the customer with an explanation of their work. It's so easy and is worth so much. You'll be requested back by this customer, which makes you valuable to the company. If hard times come, the technician that gets along with the customers will be assured of a job. There's no substitute for technical knowledge, but good customer relations is a close second."

# Troubleshooting a Water-Cooled Condenser

Bob heads out to another call when the dispatcher notifies him that a motel owner has called to report that the 50-ton chiller that serves his motel is shutting off and the high-pressure control must be reset to get it to run again. Then it only runs for a short while before it shuts off again.

Bob is a little unsure of himself as he arrives at the job because it's a new client for him and he has never worked on a water-cooled unit before. He has been thinking on the way to the job about what he should do first. Since it's shutting down because of high pressure, he knows that he should check the part of the system that rejects the heat—the water cooling tower.

Bob goes to the equipment room, where the chiller is located, and finds the maintenance man standing there looking like he needs help. The maintenance man explains that he doesn't know anything about refrigeration equipment; he just pushed some buttons on the unit and when he pushed the high-pressure button it started up. He reaches for the button again to restart the compressor and Bob says, "Let me look around for a minute before we restart it. There may be something obvious that's keeping it from operating."

Bob looks around the equipment room and identifies the condenser water pump; it is running and seems normal. He then goes to the cooling tower for a look. Water is flowing over the cooling tower and it appears normal. A look into the cooling tower sump shows the water to be clean-looking. All of this seems good, so Bob decides to install gauges on the chiller and start it up to see what happens.

BTU Buddy appears at this time and asks, "What do you expect the gauges to read when you start the compressor?"

Bob says, "I'm not sure. I thought I would start it up and see what they read."

BTU Buddy says, "You need to have some expectation or guideline before you start it up. Here are some numbers that you should commit to memory. The chiller is an R-22 machine. The suction pressure should operate at about 70 psig corresponding to 40°F evaporator when the chilled water leaving is 45°F and the entering water is 55°F. These are typical design conditions for a water chiller. The head pressure should be about 210 psig when the cooling tower is furnishing 85°F water to the condenser. The entering chilled water is now running 80°F so the chiller really has a load on it. You can expect the suction pressure to be high and the head pressure to be lower than normal because the cooling tower is still operating. The cooling tower water is 5°F warmer than the entering chilled water.

"Here are some facts that you should know about water-cooled condensers.

1. The cooling tower should provide water that is about 7°F warmer than the outdoor wet bulb, so a 78°F wet bulb should yield 85°F water to the condenser. This is called the cooling tower 'approach temperature.' The cooling tower can reach an approach temperature of 7°F. There are towers with closer approach temperatures, but 7°F is typical. The dry-bulb temperature doesn't influence the cooling tower water temperature. It's controlled by the wet-bulb temperature which is related to the evaporation rate.
2. Most condensers have a 10°F rise in water temperature from the inlet to the outlet, so 85°F inlet should mean 95°F out. Many technicians refer to this as the 'split' in temperature. Remember, this is under full load. When the water-cooled condenser is only working at half load, there will be a 5°F rise in condenser water temperature.

"There is a relationship between the exiting condenser water and the refrigerant condensing temperature of about 10°F. For example, if the refrigerant is condensing at 105°F, the exiting condenser water should be about 95°F. This is known as the condenser approach temperature." Figure 12-1 illustrates this. "This is important to know because when we

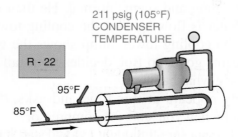

211 psig (105°F)
CONDENSER
TEMPERATURE

R - 22

95°F

85°F

**FIGURE 12-1.** These are the typical readings for a water-cooled condenser using a cooling tower.

start the compressor, we will record these conditions and see how they compare with normal conditions. If they don't make sense, we can find out why by taking the following readings:

Condenser water outlet temperature: ___°F
Condenser water inlet temperature: ___°F
Temperature difference: ___°F

A. Condensing temperature (from head pressure): ___°F
B. Condenser water out temperature: ___°F
Approach temperature (A – B): ___°F

"There is something that we should consider here. Are the two thermometers on the condenser water correct? They are permanent thermometers and have been in service for a long time. Since the chiller is not operating, both should read the same because the condenser is not adding any heat to the water. What are they reading?"

Bob looks and says, "They are reading the same, 84°F."

BTU Buddy says, "That's good. When you start the compressor, watch the head pressure and record it before it can shut down. It will probably run for a few minutes before it shuts off. Also watch both condenser water temperatures and record them."

Bob resets the high-pressure control. Everything seems fine for a few minutes, and the head pressure starts to rise. It rises to 240 psig and BTU Buddy says, "You better record your readings because it's going to shut off soon."

Bob takes down the readings and just before it shuts off, BTU Buddy says, "Feel the liquid line leaving the condenser; I'll ask you about it in a minute."

The chiller shuts off on high pressure.

Bob reviews the readings:

Condenser water outlet temperature: 90°F
Condenser water inlet temperature: 85°F
Temperature difference: 5°F

A. Condensing temperature (from head pressure): 117°F
B. Condenser water out temperature: 88°F
Approach temperature (A – B): 29°F

BTU Buddy asks, "What do you see in these readings?"
Bob says, "I'm not sure. I need help."

BTU Buddy says, "The condenser should have a 10°F difference from inlet to outlet, and it only has 5. That tells you that it's not doing much work, actually only 50% of its workload if 10°F is normal. The approach temperature is 29°F when it should be about 10°F. That tells you that the condenser is not removing enough heat. Now for the question: What did the liquid line feel like when you touched it—cool, warm, or hot?"

Bob answers, "It was much warmer than hand temperature. Why does that matter?"

"Well," BTU Buddy explains, "if it were cool, the system could have an overcharge of refrigerant and a lot of subcooling due to the cool temperature of the cooling tower water. If it were hot, that would mean that the condenser is just not taking the heat out of the refrigerant. I would say that you have a case of dirty condenser tubes. About all you can do is shut the system off, drain the condenser, and go in for a look."

Bob shuts everything down and removes the condenser head, and is surprised to see a lot of slime in the condenser tubes.

"I don't understand. The water looked good," Bob exclaims.

BTU Buddy says, "The tower may have just been drained and cleaned, but the condenser is still dirty."

Bob uses brushes to clean the condenser, then he rinses it out with fresh water. He then puts the cover back on.

The tower is filled and the water pump is turned on, and Bob is ready to start the system again. He asks, "Should I look out for anything in particular?"

BTU Buddy responds, "Now is a good time to record the previous temperatures and then see what this chiller does with a clean condenser. That is good information to have on the jobsite and in your records for this job. The next time, a different technician may be the one out here and those records will be appreciated."

Bob starts the chiller and lets it run for about 30 minutes, recording the following:

Condenser water outlet temperature: 95°F
Condenser water inlet temperature: 85°F
Temperature difference: 10°F

A. Condensing temperature (from head pressure): 105°F
B. Condenser water out temperature: 95°F
Approach temperature (A − B): 10°F

"That's perfect," BTU Buddy says.

Bob says, "The chilled water outlet temperature is 45°F, so the system is really working."

"These systems are very predictable if you take the time to understand them," BTU Buddy says. "You should take a couple of copies of these figures and put one in the file and the other in the control panel of the chiller. This may save you some time in the future. Tell the maintenance man where they are, and someday, if he calls you again with a problem, you can ask him for the temperature readings over the phone to have some idea of what the problem is."

Bob shows the paper to the maintenance man who clearly appreciates what Bob has done.

BTU Buddy says as they are riding away, "You've made another customer happy. Just keep building happy customer files and you'll always be in demand."

# Examining a Cooling Tower Problem

A manager of a group of movie theaters calls and tells the dispatcher that the air conditioning system is not working at a shopping mall that has six theaters. The manager explains that it was working last night, but is not working this morning.

Bob is sent to handle the job. He arrives, and since he has never been there before, the manager takes him to the equipment room. He finds a water-cooled chiller that looks to be supplying about 50 tons of air conditioning. He tells the manager that he wants to look around to see what the problem could be.

Bob looks over the chiller and discovers that the high-pressure control has the chiller off. He can tell because it has a red plunger that's sticking out of the control. He starts to push it in when BTU Buddy appears and says, "Why don't you look over the cooling tower before you reset it?"

Bob looks at the cooling tower, which appears to be fine; good, clean-looking water is going over the top. So he goes back into the equipment room and finds that the temperature of the water is cool. There are no working thermometers in the condenser water lines, but the pipe is cool.

BTU Buddy suggests that he take a temperature tester and place one lead at the top of the tower where the water comes out, and one in the tower basin where he can check the temperature. The tower is just behind the equipment room on the ground. He also suggests that Bob put gauges on the compressor.

Bob gets everything set up and BTU Buddy says, "Remember the last water-cooled call and get ready to write down the performance information."

Condenser water outlet temperature: ___ °F
Condenser water inlet temperature: ___ °F
Temperature difference: ___ °F

A. Condensing temperature (from head pressure): ___ °F
B. Condenser water out temperature: ___ °F
Approach temperature (A – B): ___ °F

Bob pushes the reset button and the chiller starts up. It seems normal. The chilled water begins to lower.

Bob is looking at the gauges when the head pressure begins to slowly rise. He waits a few minutes and records the following:

Condenser water outlet temperature: 100°F
Condenser water inlet temperature: 95°F
Temperature difference: 5°F

A. Condensing temperature (from head pressure): 110°F
B. Condenser water out temperature: 100°F
Approach temperature (A – B): 10°F

About then, the chiller shuts down because of high head pressure.

BTU Buddy asks, "What do you know from looking at the readings that you took?"

Bob studies them for a minute and says, "The leaving-condenser-water to refrigerant-condensing-temperature (approach) looks normal; the condenser is doing its job. The cooling tower water is too hot coming back from the tower basin. The fan on the cooling tower is running, so I would guess that the tower is not doing its job for some reason. I can't imagine why."

BTU Buddy says, "Let's go and take a look."

Bob inspects the tower and says, "I don't see anything wrong."

BTU Buddy says, "Let's review how a tower works. The object is to allow water to evaporate at the correct rate in order to lower the temperature of the remaining water that returns to the condenser. The tower uses a pan in the top with calibrated holes in it to spread the water over the entire tower area," Figure 13-1. "It has a fan to speed the rate of evaporation in a smaller space. The tower also has slats, called 'fill material,' that are supposed to spread the water out to provide more surface area for better evaporation," Figure 13-2. "With that in mind, look the tower over very carefully and see what you can find."

Bob examines the tower for a few minutes and says, "The upper tower basin has a big hole in one end. It seems that too much water is flowing down that hole. This would cause it to flow past the fill material," Figure 13-3. "Is that what I'm looking for?"

BTU Buddy says, "Yes, that is the problem. This is a very old system and a rust hole has caused the water distribution to not flow over the

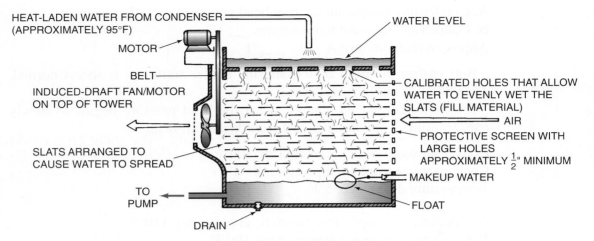

HEAT-LADEN WATER FROM CONDENSER
(APPROXIMATELY 95°F)

MOTOR

BELT

INDUCED-DRAFT FAN/MOTOR
ON TOP OF TOWER

SLATS ARRANGED TO
CAUSE WATER TO SPREAD

TO
PUMP

DRAIN

WATER LEVEL

CALIBRATED HOLES THAT ALLOW
WATER TO EVENLY WET THE
SLATS (FILL MATERIAL)

AIR

PROTECTIVE SCREEN WITH
LARGE HOLES
APPROXIMATELY $\frac{1}{2}$" MINIMUM

MAKEUP WATER

FLOAT

**FIGURE 13-1.** A typical induced-draft cooling tower.

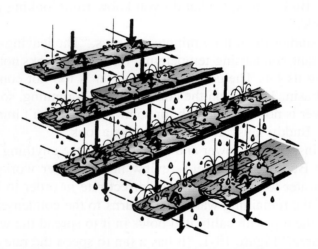

**FIGURE 13-2.** This is an example of how the fill material spreads the water out for a better evaporation rate.

tower fill like it should. This has been coming on for a long time and no one noticed it."

Bob asks, "Well, what can we do about it?"

BTU Buddy says, "Here are the options: We know that the tower must be replaced. It will take several weeks to get a replacement tower and the theater will lose a lot of money being shut down for that long. We could suggest to the owners that they convert this chiller to an air-cooled condenser, which would be more reliable, but that would also take several weeks to arrive. Meanwhile, no matter which choice the owners make, we should patch this tower until a replacement can be located."

"How in the world can we patch a rusty mess like that tower basin?" asks Bob.

"Well, there are two options," says BTU Buddy. "You could get a welder over to make a temporary patch that would probably last the rest of the season, or you could make a plywood patch over a portion of the tower and drill some holes of the same size. That also would probably last the rest of the season."

Bob says, "Why don't we bring the manager out and show her what the problem is. She may want to talk to the owner about what to do."

The manager then comes out and looks at the tower hole, and asks Bob to let her call the owner. She calls the owner on her cell phone and asks Bob to explain the options. When Bob is through, she talks to the owner again and verifies what Bob has explained to him.

The owner tells the manager to have Bob make a plywood temporary fix and to get a price on both the air-cooled and water-cooled options. He explains to her that he's had similar problems before and changed out to air-cooled systems and that it was a good experience.

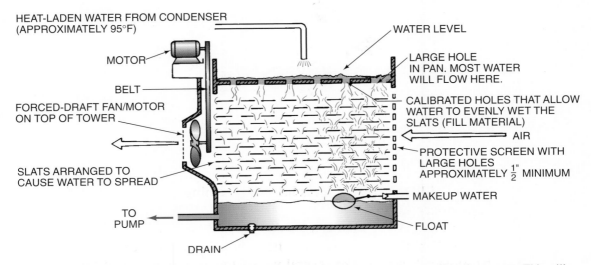

**FIGURE 13-3.** This cooling tower has a large hole in the pan that allows too much water flow in one spot. This will prevent the tower from reducing the water temperature and will cause high head pressure.

Bob gets to work on the plywood patch and installs it by just fastening it to the tower basin with long sheet metal screws. He starts the chiller up again, and after it has run for about 30 minutes, he records the following:

Condenser water outlet temperature: 96°F
Condenser water inlet temperature: 86°F
Temperature difference: 10°F

A. Condensing temperature (from head pressure): 105°F
B. Condenser water out temperature: 96°F
Approach temperature (A − B): 9°F

Bob says, "Boy, that was a quick recovery for the theater. They really appreciated being able to have cooling so soon."

BTU Buddy says, "You've made another customer happy with your service. Your company is really beginning to appreciate your worth. I understand you got a raise; did it help?"

"Yes," says Bob, "the company owner called me in and said that he'd had several calls from customers who were satisfied with my work. They said that I'd kept them informed about progress, and that they felt they were receiving quality service for their money. The owner was very pleased."

# Compressor with Refrigerant Flooding Back

It's a sweltering August day when Bob receives a call from the dispatcher telling him that an office building is not cooling properly and needs to be checked. This is a new customer for the company, and their system is about 7 years old.

Bob arrives at the office and goes to the woman who manages the building and handles all mechanical calls. She says, "The building is not holding the temperature conditions that it normally does. It also feels very humid. It started yesterday and it's still uncomfortable."

Bob asks to see how she knows what the building temperature is, and she shows him a quality thermometer that she uses to take around the building when there are temperature complaints. Bob says, "You really look out for the building; that's good."

The manager then leads Bob to the equipment room where the air handler is located, and leaves him to investigate the problem himself.

He starts by feeling the lines to the air handler. He discovers that the unit must be running and cooling because the suction line is very cold. In fact, it seems too cold. He then goes to the condensing unit, which is out back behind the building. He can tell that the unit is running and putting out heat because the air leaving the fans is really warm. He decides to remove the compressor compartment door, and what he sees surprises him. The compressor is very large and is sweating all over the motor housing and the crankcase, Figure 14-1. It's obvious that it's running and cooling; the question is why is it not cooling the building?

About this time, BTU Buddy appears and says, "What's the matter, Bob?"

Bob explains what he knows; that the system shows all signs of cooling except that it's not holding the building temperature as it normally does.

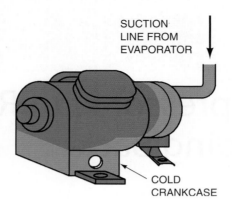

SUCTION
LINE FROM
EVAPORATOR

COLD
CRANKCASE

**FIGURE 14-1.** This compressor has a cold crankcase because of liquid refrigerant flooding back to the motor housing and crankcase.

BTU Buddy suggests, "Put your gauges on the system and I think you'll find the answer."

Bob fastens gauges to the system and asks BTU Buddy, "Why is the suction pressure so high? I would expect it to be about 70 psig because the refrigerant is R-22. But look at this: it's 79 psig. The suction line is so cold that you can hardly hold it."

BTU Buddy says, "Get your temperature tester and see what the superheat is at the compressor."

Bob fastens a temperature lead to the suction line and says, "The superheat is 0. The suction pressure is 79 psig which means the refrigerant is boiling at 47°F and the suction line is reading 47°F. Refrigerant is flooding back to the compressor but it's not knocking or making any excess noise like I thought it should."

BTU Buddy says, "Find out why it's flooding back and cure the problem, then we'll discuss what you found. Let's protect the compressor first. What would be a good reason for it to flood back?"

Bob says, "The expansion valve isn't controlling like it should. I'll take a look at it."

## CHECKING THE EXPANSION VALVE

Bob goes to the equipment room to examine the expansion valve. It shows no signs of anyone having removed the cover to the expansion valve adjustment, so he starts looking at how the expansion valve bulb was mounted. He finds that it had been mounted on the suction line using nylon tie-wraps which have both broken from age. The bulb was very loosely insulated.

BTU Buddy says, "Shut the king valve at the condenser, and pump the system down into the condenser while you make the repair, to prevent any more flooding of the compressor."

Bob pumps the system down into the condenser and asks, "Now what?"

BTU Buddy explains, "Now is the time for an explanation of how an expansion valve bulb should be mounted. The object of mounting the expansion valve bulb correctly is to accurately measure the temperature of the vapor in the suction line." BTU Buddy goes on to outline the 5 factors that enter into locating the bulb:

1. It should be mounted on a horizontal portion of the line.
2. The bulb should be mounted at different places on the circumference of the line, depending on the line size. On a small line, it should be mounted on top. The larger the line, the closer to the bottom it should be mounted, but *never on the bottom,* because oil is also circulating and can act as an insulator. Oil will travel on the bottom of the cold suction line.
3. The bulb should be located before the connection for the external equalizer line, because if the valve had an internal leak, it would give a false reading and starve the evaporator coil of refrigerant, see Figure 14-2.
4. It should be firmly mounted on the correct position on the line, see Figure 14-3.
5. The bulb and line should be well insulated with no air passing over the bulb.

"If any of these steps are omitted," BTU Buddy continues, "good control of the superheat may not be possible. Which of these 5 items has been violated?"

Bob looks at the bulb with the checklist in mind and says, "The bulb is not fastened tightly and is not properly insulated. How has it operated for years without a problem?"

BTU Buddy explains, "This has been an advancing problem. First, one of the straps failed and the system was beginning to lose control. Then the second strap failed and pushed the bulb insulation up. That caused air that was blowing out of the side of the air handler, where the suction line goes through the cabinet, to begin blowing under the insulation, and the bulb lost control. It probably hasn't done damage yet, but it must be repaired now or damage to the compressor will occur."

Bob asks, "Why was the system not cooling and dehumidifying? The coil sure was cold."

BTU Buddy explains, "The design temperature for this coil would typically be 40°F and it was operating at 47°F. This would raise the outlet air

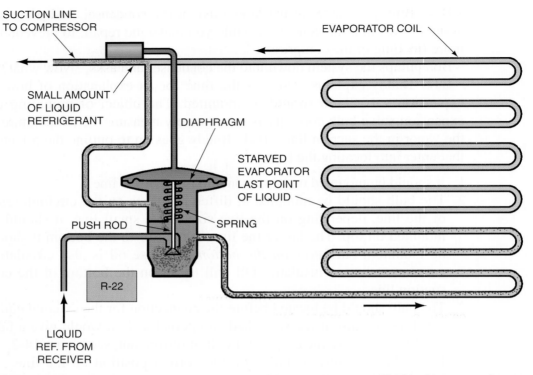

**FIGURE 14-2.** This expansion valve has an internal leak causing a small amount of liquid refrigerant to migrate to the suction line. The expansion valve is mounted after the external equalizer connection so this liquid refrigerant is causing the bulb to believe the evaporator is full of refrigerant, and it starves the evaporator.

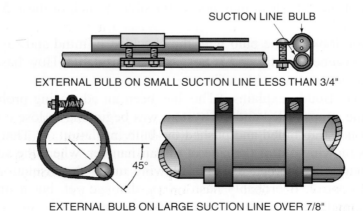

**FIGURE 14-3.** These are the best positions for mounting the sensing bulb for a thermostatic expansion valve.

temperature by 7°F, and would reduce the moisture removal ability because of the rise in coil temperature."

By this time, Bob has fastened the bulb down on the line correctly, using two copper straps, and has insulated it correctly and tightly so that air cannot pass over the bulb. It's now time to start the system again.

Bob turns on the system and, as it runs, the suction line warms up to about 14°F of superheat at the compressor. The compressor stops sweating and the suction pressure decreases to 72 psig and is still falling.

"Now," Bob asks, "why wasn't the compressor making a noise?"

BTU Buddy responds, "The refrigerant returning to the compressor was what we call 'wet vapor' with a small amount of liquid in it. It has 0°F superheat, but not enough liquid to reach the compressor cylinders. The liquid was boiled to a vapor in the motor housing and the crankcase of the compressor. The long-term damage of this is reduced lubrication to the compressor. If you can, imagine a comparison to your car: if it should use 40-weight oil and you've diluted the oil with 50% kerosene, the oil would not maintain the oil film between all moving parts. It wouldn't cause immediate damage, but the more you operated it, the more damage it would cause. The reason for pumping the system down was to prevent excess flooding while you were working with the expansion valve bulb, and to pump any liquid refrigerant out of the compressor for when we restarted it after the bulb repair."

Bob asks, "Why did the suction line feel so cold? It certainly didn't feel like 47°F. It seemed much colder."

BTU Buddy explains, "When there is liquid refrigerant in the line, it will take the heat out of your hand so fast that it seems to hurt your hand. It's still 47°F, but the heat is removed faster and it feels colder. Also, the heat from your hand will not quickly cause the suction line to rise in temperature like it would with vapor only in the line."

Now the system is operating correctly and Bob goes to visit the manager and explains the problem and the repair. He talks to her about a service contract for the building that would involve regular filter changes, motor lubrications, and performance checks for cooling in the spring and heating in the fall. He gives her a written proposal from his company.

She thanks Bob for a professional job well-done.

Riding away, BTU Buddy says, "You have garnered another customer for your company. You've added to their business, which makes you valuable to them."

# Compressor with Low Oil Level

The first cool weather of fall has prompted a service call to a commercial building that has a 100-ton split system with a single compressor. There are three evaporators, one on each floor, and an air-cooled condenser in the parking lot behind the building. The building maintenance man called and reported that the oil level was low in the compressor. It was way below the sight glass where it had normally been, so he shut the system off.

Bob arrives about 10:00 a.m. and drives around back where the equipment room door is standing open, with the maintenance man waiting for him. It's about 65°F outside, but the maintenance man says the inside of the building is getting warm and complaints are coming in. He explains, "The oil level is real low in the sight glass, but the compressor was still running and cooling. It had been on all night, only shutting off when the thermostat was satisfied. It was running when I came in this morning and I noticed the oil level so I shut it off. I have a plumbing leak that I have to tend to now. I'll be back in about an hour."

Bob uses a small penlight to look in the oil sight glass, and he sees that the oil level is very low. He stands there for a minute, thinking, and starts for his truck when BTU Buddy appears.

BTU Buddy asks Bob, "What's your plan?"

"I'm going to get gauges to put on the compressor, oil to add to the oil sump, and a pump to pump it into the sump," Bob explains.

BTU Buddy then says, "I believe that the low oil level is a normal function of this system this time of year. Let me explain. This compressor has cylinder unloaders that allow it to run at reduced load. Do you remember those terms from school?"

Bob says, "Yes, I remember them, but I can't say that I fully understand them. It seems that my schooling made me familiar with a lot of procedures, but I don't truly understand them until I've used them in the field. Would you mind reviewing the term 'cylinder unloading'?"

## CYLINDER UNLOADING

BTU Buddy explains, "This is a 100-ton system. The only times that it runs at full load are in the middle of the summer on the very hottest days, or when the system has been off long enough for the building to get warm, and then probably for only short periods of time. This compressor has eight cylinders and is rated at 100 tons, that's 12.5 tons per cylinder. Cylinder unloading allows a certain number of cylinders to unload, or stop pumping, refrigerant from the low-pressure side to the high-pressure side when the load is reduced on the system.

"It works like this: Each evaporator will have multiple coils that are solenoid valve controlled. For example, the building has three floors and three coils. The fan and evaporator coils for the first floor are right over there. Notice that the coil has two expansion valves and each has a solenoid valve before the expansion valve. This system probably has a return air thermostat that controls the solenoids. As the return air thermostat starts satisfying, it will shut off the top solenoid and stop refrigerant flow. This reduces the capacity to one-half capacity for that coil. What do you think that would do to the suction pressure at the compressor?"

Bob says, "It would go down because of reduced evaporator space."

BTU Buddy says, "That's correct. The compressor has a pressure control inside it that will stop part of the cylinders from pumping when the suction pressure drops, reducing the capacity and the power consumption. Everybody wins on that one. That's a simple process that works for all three floors of this building. A very important thing to remember is that a 100-ton (approximately 100 horsepower) compressor is not stopping and starting—it continues to run. Starting a compressor is harder on the bearing surfaces than keeping it running with the correct oil pressure. We'll get to that in a minute. The point to remember is that this compressor has capacity control. It actually has three stages of unloading so it can be a 100-ton, 75-ton, 50-ton, or 25-ton compressor, with the power reduced to the respective tonnage level it's operating at. Now, for the important statement about this job: **This compressor operated at its lowest capacity for several hours last night and this morning before the maintenance man discovered the low oil level**."

Bob asks, "What does that have to do with low oil level?"

BTU Buddy then asks Bob, "Do you see any oil on the floor around the compressor?"

Bob says, "No, but I haven't looked at all of the piping for oil."

BTU Buddy says, "Why don't you follow the piping and see if you find any oil? Meanwhile, I'm going to sit here and wait for you."

Bob returns in a few minutes and BTU Buddy says, "You didn't find any oil, did you?"

Bob says, "No. Now I'm really confused. There are probably at least two quarts of oil somewhere. What do we do next?"

BTU Buddy says, "If it's not out of the system, we can only assume that it's still in the system."

Bob asks, "Well, how do we find it? I think it would be risky to start the system up with really low oil."

BTU Buddy then asks, "Does the oil always stay in the compressor?"

Bob says, "I thought so."

BTU Buddy then explains, "All reciprocating compressors pump a certain amount of oil as a normal function of lubricating. The cylinders are lubricated with oil and when the piston rises in the cylinder, some of the oil is wiped off and pushed into the discharge line. Once oil leaves the compressor in this system, the only way back to the compressor sump is to take the whole route through the discharge line, the condenser, the expansion valve, the evaporator, and back down the suction line to the compressor, where it migrates to the sump. That's a long route. What do you think causes it to move through the system?"

Bob says, "When I studied pipe sizing, it was explained that refrigerant velocity moves the oil."

"Exactly," says BTU Buddy. "Now, what happens to refrigerant velocity when it's pumping less refrigerant through the same lines?"

Bob says, "The velocity would be less, so the oil return would be slower."

"Now you have it," says BTU Buddy. "Let's take the precaution of installing gauges on the system before we start it up. Get two sets of gauges from your truck and we'll monitor the oil pressure and the high- and low-side pressures when we start it up. Bring an ammeter so we can tell what capacity we're operating at."

Bob installs the high- and low-side pressure gauges and turns to BTU Buddy and asks, "How do we install the oil pressure gauge?"

BTU Buddy shows Bob the oil test port for the compressor and says, "Fasten the high-side gauge from the extra set of gauges here."

BTU Buddy then explains, "This system has a very reliable oil pressure shut off control that monitors the oil pressure and will only allow the compressor to run for about 90 seconds in case of low oil pressure,"

Figure 15-1. He continues, "Notice that it is manual reset and that the maintenance man didn't mention resetting any controls. The actual oil pump picks up the oil in the very bottom of the sump, so the oil level can be low and there will still be oil pressure for full lubrication," Figure 15-2. "We know this compressor has oil in the bottom of the sump because we can see it. Let's start the compressor and monitor the oil pressure. Do you understand the term 'net oil pressure'?"

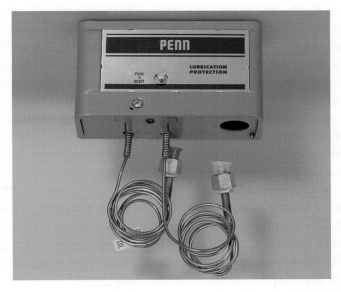

**FIGURE 15-1.** This is an oil safety control. Notice that it has two connections. One will be fastened to the suction side of the system and the other to the oil pump.

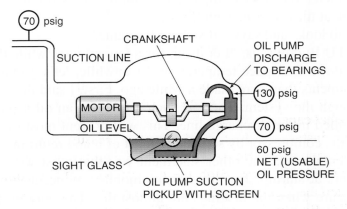

**FIGURE 15-2.** This oil sump diagram shows that the oil pickup point for the oil sump is low in the compressor sump. Also note the net oil pressure calculation. It is also different from the one in our example problem. Compressors have different net oil pressures.

Bob says, "Again, I know the term but I really don't understand how to apply it."

## NET OIL PRESSURE

BTU Buddy explains, "Oil pressure is a floating pressure that depends on the suction pressure. The oil pump inlet pressure is always the same as the suction pressure, which can vary from the off cycle to the run cycle. For an R-22 system like this, the system-off pressure may be 140 psig and when the system is running, the suction pressure would be about 70 psig," Figure 15-2. "That's what I mean by floating, so we'll use the suction pressure as the oil pump inlet pressure and the oil pump discharge gauge reading as the oil pressure. We'll then subtract the suction pressure from the oil pump discharge pressure for the 'net oil pressure.' The net oil pressure may run from 20 psig to 60 psig depending on the manufacturer. With this system we can expect the net oil pressure to be about 40 psig. With the suction-pressure-while-running expected to be about 70 psig, the oil pump discharge pressure will be about 110 psig (70 plus 40 equals 110). Now start the compressor."

Bob starts the compressor with the system switch. It starts up and sounds good. The oil pressure discharge gauge is reading 115 psig with the suction pressure reading 75 psig, so the net oil pressure is 40 psig.

BTU Buddy says, "There is plenty of oil pressure, so the compressor is safe in that respect. Now, check the amperage."

Bob reads the amperage and it's 110 amps.

BTU Buddy says, "The name plate full-load amperage is 115 amps and the meter reads 110 amps. The compressor is running at near full-load. Look at the oil in the sight glass."

Bob looks and says, "It's about the same."

BTU Buddy says, "Let it run for a few minutes while the system is pulling down the space temperature. Meanwhile, let's go back to a couple of statements that were made a while ago. First, I said that the oil must return through the system. There is such a thing as an oil separator that may be installed in the discharge line. This will prevent most of the oil from leaving the compressor by separating most of it and returning it to the compressor crankcase. The other statement was that most wear on a compressor occurs at startup. Just like an automobile engine, it starts up with no oil pressure when the bearings are almost dry. The only oil in the bearings is what did not drip out during the off cycle. Let's meet for lunch tomorrow for a review of what you've come across today."

After about 15 minutes, the compressor makes a tone change and Bob asks, "What was that?"

BTU Buddy explains, "The compressor just unloaded two cylinders. It's now running at 75 tons. What does the ammeter read?"

Bob looks and says, "The amperage has gone down to 85 amps."

BTU Buddy then says, "Look at the oil in the sight glass."

Bob looks and says, "It's half full. It's a good thing I didn't add oil, or there would be too much now."

BTU Buddy says, "You'd be surprised at how many technicians would add and then remove oil when they only needed to load the system up to return the oil that is out in the system. You should explain to the maintenance man what you did and what he should do in case it happens again."

The maintenance man returns in a few minutes and Bob explains that nothing abnormal had happened, that the system had been running under reduced load for many hours and that the oil would have returned to its normal level in the sight glass when the load picked up again. He also explains that if it happens again, he can shut the system down until a load builds up, and then restart the system and the oil should return. He continues that he verified with gauges that the compressor had plenty of oil pressure and lubrication, and that all was well.

The maintenance man says, "I thank you for explaining all of that to me. It's all very interesting."

Bob then says, "You can take classes at the local technical school that will give you a great education on these kinds of systems. Your employer would probably pay for any completed classes, which sounds like a win-win situation to me."

BTU Buddy notes as they are driving away, "That was a nice added benefit that you suggested to the maintenance man. Many people like him could raise their stock in life with more specialized education. Our profession needs more interested people. It's a wide-open field."

# A Training Lunch with BTU Buddy

Bob and BTU Buddy meet for a training lunch after the previous day's low-oil-in-the-sight-glass service call.

Bob starts the conversation by saying, "The service call on that 100-ton compressor that didn't show any oil in the sight glass was really different yesterday. It seems like there is a lot to learn in this HVACR field."

BTU Buddy says, "Yes, there is a lot to learn and we're only been working with the HVAC side, no refrigeration. That's another subject altogether. There are a lot of similarities, but a lot of differences."

Bob says, "Let me learn the HVAC part before trying to explain the commercial refrigeration side of the business."

BTU Buddy says, "Good idea. Most technicians work on either one or the other. Heating and cooling technicians are rarely called on to work on refrigeration, and refrigeration technicians are rarely called on to do heating or cooling. The common thread is the refrigeration cycle used for air conditioning and refrigeration. Actually, air conditioning uses high temperature refrigeration to cool air and water.

"Now let's get down to business. We were involved with oil and oil return with the service call yesterday. Let's start by saying that there are two fluids circulating in every refrigeration system; refrigerant and oil. They depend on each other for the proper circulation. Refrigerant oil is primarily used to lubricate the moving parts inside the compressor. There is actually supposed to be a thin film of the proper weight of oil between all moving parts. In a reciprocating compressor there are a lot of moving parts, but there are fewer in scroll compressors and rotary compressors. The same oil theory is true for them, though. Without oil, they won't run for long. Notice that I said a *thin film* of the proper weight oil. Like a car, there are different weights of oil for different applications.

"Also, recall the service call that we had awhile back where liquid refrigerant was slowly moving into the compressor crankcase and causing the oil to be thin. Liquid refrigerant and oil make for poor lubrication. The compressor oil crankcase should always be warm; then you know there is no liquid refrigerant present."

Bob asks, "Why don't you change the oil in a compressor like you do in a car?"

BTU Buddy explains, "A car has an internal combustion engine that gives off soot or carbon as a natural by-product of the combustion. A compressor and refrigeration system are supposed to be assembled under very clean conditions, and what little contamination may be left in the system is supposed to be removed by the filter drier. Clean field-assembly procedures are essential. Air conditioning technicians have been known in the past for careless assembly. A little moisture will normally never be noticed. This is not true for refrigeration technicians. A circulating drop of water will stop a system by freezing at the expansion device, because most of them operate below freezing. The compressor, evaporator, condenser, and all other components are shipped to the job clean. When assembled correctly in the field, it is a clean system. The lubricating oil is in the system for life unless the compressor is changed."

## OIL CIRCULATION

BTU Buddy says, "Let's look at how the oil circulates in the system. As we talked about yesterday, the velocity of the refrigerant moves the oil along in the bottom of the piping, through the coils and interconnected piping. The coils are normally not a problem as they're designed at the factory for refrigerant and oil. The problem can be the field piping. This piping must be sized correctly or the oil will leave the compressor and not return. The compressor will run for short periods of time with low or no oil pressure, but not for long. Interconnected pipe sizing is a long lesson of its own that we don't have time for today. There is a delicate balance between piping that is oversized and undersized. Oversized piping will not have enough velocity to return oil, and undersized piping will cause too much pressure drop and cause the system to be inefficient. It's fairly easy to move oil downhill, but it becomes harder to move it uphill, as when the compressor is above the evaporator. The warmer the refrigerant, the easier it is to move oil. This is a difference in a refrigeration system with suction lines that may operate below 0°F, and air conditioning systems that operate at around 50°F.

"The system we worked on yesterday had all of the evaporators above the compressor. The problem was that the compressor operated at multiple capacities because of cylinder unloading. That compressor could operate at 100, 75, 50, and 25 tons of capacity, and all with the same single refrigerant line. At the low capacity of 25 tons, the line velocities were reduced to the point that the oil didn't return quickly. When the system was shut down for awhile and restarted, it operated at 100 tons for a few minutes and the refrigerant velocity at full load returned the oil.

"Now suppose that the oil had dropped to a point that no oil pressure was available. I told you yesterday that the system had a very reliable oil pressure control. Do you know how those work?"

Bob says, "I would assume that it is a typical pressure control; when it senses pressure it would have closed contacts, and when there's no pressure, the contacts would open."

BTU Buddy responds, "That is true up to a point. It has an internal pressure control that operates much like you said. Now the question is, at what pressure would you set the control? Most compressors must have a minimum of 10 psig of net oil pressure for lubrication. Remember, the oil pump inlet pressure is the same as the suction pressure and it is a variable. We can't just set the control to shut off at 10 psig. We must use the control to measure the pressure difference across the oil pump. This brings up another problem: How do you measure pressure difference?

"I like to explain that pressure difference is like a seesaw. What happens if a 50-pound kid gets on one end of the seesaw and a 75-pound kid gets on the other? Of course the seesaw lowers on the heavy end because there is a difference in pressure. An oil safety control operates the same way. You can look at one of the older-style controls and see that it has a diaphragm on the top of the control and a diaphragm on the bottom of the control; they're linked inside to counterbalance each other. The suction pressure is on one side and the oil pump discharge pressure is on the other," Figure 16-1. A simplified drawing of how these controls work, whether they are the new electronic versions or the older electromechanical types, can be seen in Figure 16-2.

"Here's how it works. When the compressor is off, there is no net oil pressure. The compressor starts up with whatever oil is still sitting in the bearings. It takes a few seconds for the compressor to establish pressure, so the control has a built-in time delay. That is the resistor that is built into the control. When the compressor is started, power is supplied to the oil pressure safety control, and the resistor becomes warm. If it is left energized for about 90 seconds, it will heat the bimetal switch above it, open the

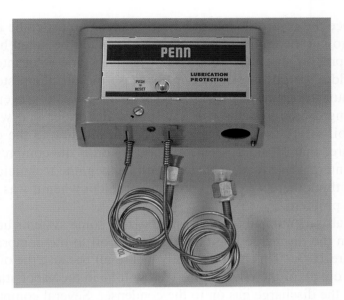

**FIGURE 16-1.** This is an oil safety control. Notice that it has two connections. One will be fastened to the suction side of the system and the other to the oil pump.

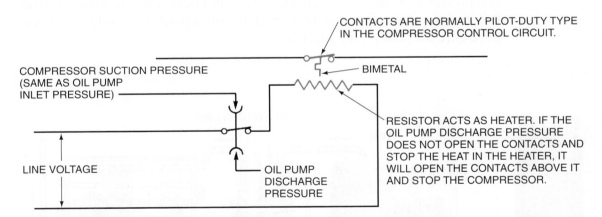

**FIGURE 16-2.** This diagram shows how the oil safety control works. Note the resistor and heater circuit.

contacts, and stop the compressor. The object here is to produce oil pressure within the 90-second period and open the circuit to the heater so the compressor will continue to operate."

Bob asks, "What would happen if the compressor were running and lost oil pressure? Would the heater circuit be energized and stop the compressor?"

BTU Buddy answers, "Yes, after 90 seconds or whatever time delay the manufacturer uses. The 90-second delay prevents nuisance shutdowns.

Notice the control is manual reset. Someone must come and push the button to get the compressor to restart. It should be noted here that should a compressor shut down because of an automatic reset internal overload, it would also fail on oil pressure because the oil pump stops pumping oil. Manufacturers have put other controls in place to prevent this from happening. Some systems use a current transformer that will only energize the oil safety control when there is current flow to the compressor. This ensures that oil safety control won't go on if the compressor stops due to the internal overload. If that were to happen, the technician would mistakenly think there was an oil pressure problem when it was actually an overload problem.

"I also mentioned to you yesterday that some systems have an oil separator in the discharge line that will return oil to the compressor crankcase," Figure 16-3. "This is not actually a cure-all because the oil separator won't take out all of the oil, as some of it will be in the vapor state and will move with the discharge gas on to the condenser. Several companies have different technologies and have different efficiencies for oil separators. For the system we worked on yesterday, the oil would probably never leave the compressor if it had an oil separator. It would have some runtimes at near full load to return oil that went through the oil separator. An oil separator is not a cure for poor piping practices."

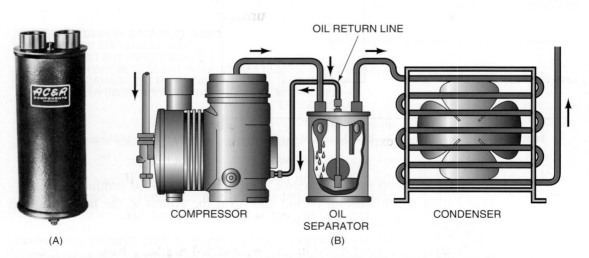

OIL RETURN LINE

(A)  COMPRESSOR  OIL SEPARATOR  (B)  CONDENSER

**FIGURE 16-3.** An oil separator (A) removes most of the oil from the hot discharge gas and (B) returns it to the compressor crankcase when the float rises.

## GOOD PIPING PRACTICES

Bob says, "Explain some tips for good piping practices."

BTU Buddy notes, "Well, we've determined that velocity of the refrigerant is the key to moving the oil. Let's first talk about refrigerant gas lines. The lines should have a minimum velocity of about 500 fpm (feet per minute) and a maximum of 4,000 fpm. With a velocity of less than 500 fpm, oil will not return. A velocity of over 4,000 fpm will cause inefficiencies and cause the piping to make excessive noise.

"Horizontal lines can be sized larger because the oil will be easier to move. All horizontal lines must be pitched downhill 1/2 inch per 10 feet in the direction of flow to encourage the oil to move.

"Vertical lines are a different story. For a compressor that doesn't have the capability to unload, a single pipe may be used for vertical lift. It must have a trap about every 10 feet where oil will gather and cause an oil plug in the pipe. When this plug restricts the flow enough, the pressure drop across the trap will pull the oil puddle up the riser.

"Systems with variable capacity may need to have two suction lines to carry the refrigerant and oil up the rise. For example, the 100-ton compressor that would run at 100 tons down to 25 tons may have a 75-ton line and a 25-ton line that are piped in such a way that, at low load, the trap in the large line will gather oil and force the oil up the small 25-ton line or riser. This piping system is sized normally by the design engineer for the system. There are texts that explain how to size these lines. Copeland©, Carrier©, and Trane© are manufacturers that furnish these texts. Normally, the service technician would use their tables to check a system where trouble is suspected."

"Wow," says Bob, "this gets deeper by the day."

"Yes it does," says BTU Buddy. "Fortunately, you can learn it a little at a time, as you go. It would be much more difficult for a technician to learn it all and then go into the field. The best way to learn is to travel in a truck with an experienced technician for some time before going out on your own, like you did. That allows you to start out on jobs that are within your capacity. Every technician can have a BTU Buddy if they'll just look around. It may be a textbook that they refer to, a friend at a supply house, or another service technician; not to mention night classes and factory schools."

Bob says, "Thanks for the luncheon lesson. I'm glad I have you to consult with."

# A Compressor with an Internal Leak

Bob gets a call from the dispatcher that a drug store is having trouble with its cooling system. It seems that the system had been working well until yesterday afternoon when it began to get warm in the store. It's a 7.5-ton unit with a reciprocating compressor using R-22.

Bob arrives at the store at about 10:00 a.m. and goes to the manager. She explains, "The system is about 9 years old and has never had any major repairs, just a new condenser fan motor about 3 years ago. We change the filters on a regular basis. The outdoor unit is on the roof."

Bob leans his ladder against the building to access the roof. He loads some tools in a canvas bag and ties a rope to it so he can climb to the roof and pull the bag up once he is there. When he and his tools are at the unit, he notices that the fan is running, but not the compressor. He figures that this is a classic case of low charge, with a hot compressor shutting off because of internal overload. He removes the compressor compartment door and checks the compressor, which is too hot to touch.

He attaches gauges to the unit and uses the rope to pull up a cylinder of refrigerant. He then pulls up a water hose to cool the compressor, Figure 17-1. Bob shuts off the electrical disconnect to the unit and starts a trickle of water over the compressor to cool it. He lets the water run over the compressor for about 15 minutes, and then gets ready to add refrigerant before starting the unit.

Bob switches the power on and the compressor starts up. He watches his gauges to get ready to add refrigerant, but the suction pressure is not low (80 psig), and the head pressure is 300 psig, normal for the conditions. He decides not to add refrigerant but to observe the pressures. As he is watching the pressures, he feels the compressor. The suction gas returning is really cool, but the compressor is not cooling down. He touches the discharge line

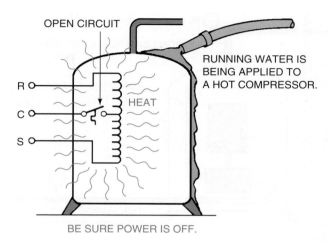

OPEN CIRCUIT

RUNNING WATER IS
BEING APPLIED TO
A HOT COMPRESSOR.

R

C

S

HEAT

BE SURE POWER IS OFF.

**FIGURE 17-1.** This method of cooling a compressor is used to get the compressor back on-line quickly. Otherwise, the technician will have to either stand around until it cools, or come back later. Both are expensive.

and it's very hot. He drips some water on it from the water hose and it boils away quickly. Bob scratches his head.

BTU Buddy appears and asks, "What seems to be the problem?"

Bob says, "I don't know. This compressor is running really hot, but everything else appears normal. There seems to be enough refrigerant returning to cool it, but it's not working."

BTU Buddy says, "The heat must be coming internally from the compressor. Get a long screwdriver and place it on the discharge line close to the compressor, and put your ear to the handle. This is a poor man's stethoscope. Then turn off the power and listen."

Bob gets everything in place and shuts the unit off with the disconnect switch. A funny look comes across his face and he says, "I hear hissing inside the compressor."

BTU Buddy says, "Feel the suction line now."

Bob says, "It's getting warm. It's not staying cool like it was. I don't get what's going on."

BTU Buddy explains, "This is a reciprocating compressor. When it shuts off, it's just like a check valve in the system. No refrigerant should flow from the high to the low side of the system during the off cycle. This compressor has an internal leak," Figure 17-2. "It cannot be repaired, so a compressor change is necessary. When the hot gas is recirculated back

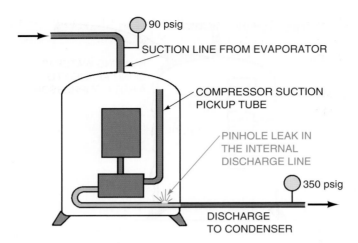

**FIGURE 17-2.** This compressor has an internal leak. Vapor is leaking from the high-pressure side to the low-pressure side causing the compressor to recompress hot gas. This will make a compressor run hot.

through the compressor, it will cause the compressor to become hot really quickly. It will then shut down because of internal overload."

"What if this had been a scroll compressor?" Bob asks.

BTU Buddy says, "Good question, Bob. A scroll compressor is a little different from a reciprocating compressor in that the internal pressures to the compressor will equalize when shut off, but there is a check valve in the discharge of the compressor that will keep hot condenser gas from backing up into the low side of the system. The funny sound that a scroll compressor makes at shut down is the high-side gas in the compressor backing up through the compressor to the low side. This is a small volume of hot gas. Not enough to warm the suction line."

"Well," says Bob, "I guess the next thing is to break the news to the store manager with a proposal to change the compressor."

BTU Buddy says, "That's all you can do for this system."

"What could have caused this internal leak?" Bob asks.

BTU Buddy says, "It could be several things:

1. It could be a vibration leak at a weak braze joint.
2. It could be a leaking discharge reed valve in the compressor head.
3. It could be the internal relief valve leaking.

If the internal relief valve is leaking, it could be because of high pressure differential between the low and the high side, such as may occur if the condenser fan is not running all of the time. After we change the compressor, we'll check the condenser fan closely."

Bob writes up a proposal to change the compressor, and then goes in and talks to the store manager. He explains, "This covers the changing of the compressor. I suspect that the fan may be the cause of the problem, but I want to check that out after the compressor is changed."

The store manager signs the proposal and asks Bob to get it done as soon as he can.

Bob assures the manager that he will work on it nonstop until the job is done.

## CHANGING THE COMPRESSOR

Bob gets the parts and arranges for a lift to help get the new compressor to the roof and the old compressor down. He also asks for a helper to move the compressor around, as it is heavy.

After he has changed the compressor and started up the unit, he starts looking over the fan motor and checks the amperage. The amperage is running a little high. Bob lets it run for a few minutes, and the fan motor shuts off. He has found the cause of the problem.

BTU Buddy then says, "It may be the fan motor or it may be the capacitor for the motor. A motor will run at high amperage if the capacitor is defective."

Bob shuts the unit off and uses his capacitor meter to check the capacitor. Sure enough, it's defective. He gets a capacitor from his truck as a replacement and installs it. He then starts the unit up again. When he checks the fan amperage, it's normal.

BTU Buddy says, "A defective capacitor can cause about a 10% increase in amperage in a fan motor. The motor will run for a while and then shut off because of internal overload. I think you've found the root cause of the compressor failure."

Bob then asks, "What caused the capacitor to fail?"

BTU Buddy explains, "It must be a random failure. Normally they last for what seems to be forever. A line surge may have damaged the internal insulation.

"You sure do ask a lot of questions, but that's good. You should be curious about failures. That means you'll understand the internals of all of the components. It will prepare you for becoming a master technician."

# A Stopped-Up Condensate Line

The dispatcher calls and tells Bob that a customer has a serious problem with a condensate leak. The unit in the attic has leaked so much that it overflowed into the attic insulation, which became heavy and caused the sheetrock in the ceiling to fall through into the upstairs den.

Bob arrives within 30 minutes and the customer meets him in the yard. The customer is fuming, as Bob's company had installed the unit about 2 years ago. Bob calms the customer down by listening to him and then says, "There must be some explanation to this. I will get to the bottom of it and give you a full report."

Bob then goes into the house to see the damage. A large piece of the sheetrock is lying across the furniture in the den. What a mess, he thinks; loose insulation is all over the place. Bob calls his company for some help cleaning up the den and removing all of the sheetrock and insulation. Meanwhile, he goes to the attic to see if he can get to the bottom of the problem. The owner has shut the unit off, so he can't really tell what the source of the water was.

As Bob approaches the unit, he finds that it is installed correctly. There is a secondary drain pan under the air handler that is supposed to catch any condensate that may run over, and route it to an outside drain, Figure 18-1. The drain is terminated at the end of the house, next to the driveway, Figure 18-2. There is an elbow turned down that would allow the secondary drain to drip water where it would be readily noticed.

As Bob gets closer to the unit, he finds out why the secondary drain was not draining. It was full of insulation; the drain pan still had water in it. Someone had added extra insulation to the attic after the air-conditioning installation, and some of that insulation had plugged the drain hole.

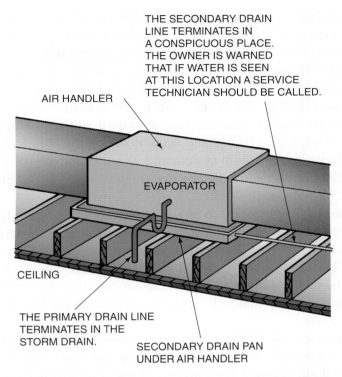

THE SECONDARY DRAIN
LINE TERMINATES IN
A CONSPICUOUS PLACE.
THE OWNER IS WARNED
THAT IF WATER IS SEEN
AT THIS LOCATION A SERVICE
TECHNICIAN SHOULD BE CALLED.

AIR HANDLER

EVAPORATOR

CEILING

THE PRIMARY DRAIN LINE
TERMINATES IN THE
STORM DRAIN.

SECONDARY DRAIN PAN
UNDER AIR HANDLER

**FIGURE 18-1.** The auxiliary drain pan should extend out past the edge of the air handler for complete protection.

THIS AUXILIARY SMALL DRAIN
FITTING TERMINATES THROUGH
THE END GABLE OF THE HOUSE.

**FIGURE 18- 2.** The auxiliary drain line should be terminated in a conspicuous location where the owner would notice it if it were to start to drip.

First things first, Bob thinks. He removes all of the insulation from inside the drain pan. It begins to drain. That takes care of that problem. Now the question is why did the primary drain pan not handle the water to the outside? He turns off the power and removes the panel to the coil, so he can observe the drain pan. It has a lot of debris in it, which looks like construction dust. Probably the people who sanded the hardwood floors or the sheetrock finishers ran the cooling system while sanding. This would generate a lot of dust that would accumulate on the coil, and much of it would run down into the drain system and would eventually stop it up.

Bob is at his truck getting his air tank when BTU Buddy appears and says, "What's next, Bob?"

Bob says, "I'm going to blow out the drain line to unplug it. It terminates over there next to the house. I'll just stick this air hose in it and give it a blast of 100-psi air. That will clear it out."

BTU Buddy says, "I should warn you that if you clear the line by blowing the dirt towards the coil and drain pan, the plug of dirt will just move to the drain pan and will again float to the drain pan drain hole and plug it up again."

Bob responds, "I knew when you showed up that I must be doing something incorrectly. You're right, but what can I do to get the dirt to the outlet of the drain system? The coil end of the drain line is too close to get the air hose in and blow out."

BTU Buddy then says, "If you can't push it, pull it."

"What do you mean by that?" Bob asks.

BTU Buddy says, "Instead of compressed air, use a shop vacuum and pull the pipe clean. Just connect the suction to the drain termination point and tape it on. Then turn on the vacuum and pull the dirt out."

Bob then says, "That's so simple. Why didn't I think of that?"

Bob connects the vacuum and turns it on, and the vacuum pulls hard for just a second. Then he hears the dirt plug come down the pipe into the vacuum with a lot of water.

"That worked great," Bob says.

## PREVENTING A PLUGGED CONDENSATE LINE

"You've solved the problem as of now," says BTU Buddy. "But there are also some things that you can do to prevent a reoccurrence:

1. The coil and pan need some cleaning, as there is more debris there.
2. The secondary drain pan installation meets the codes of this area, but you can still improve the reliability of the system by installing a drain

pan switch that will shut off the condensing unit if the water level rises. This may seem like overprotection, but it would prevent the kind of accident that just occurred. It's a low-voltage connection with all of the wires already at the air handler. All you have to do is run the yellow wire in the air handler (the wire that energizes the condensing unit) through the drain pan float switch. This is an 'add on' device that the customer will have to pay for, but it's not expensive."

Bob uses an approved detergent and cleans the coil and drain pan. The flow of detergent and water through the drain system cleans the trap and the piping. The drain system is now working perfectly.

Bob then goes to the owner and reports what he found out about the insulation. The owner says, "My brother-in-law added that insulation. He was out of work last winter and I thought anyone could do that job. He just went to a local hardware store and purchased bags of loose insulation and spread it around. I still don't understand about the dirty coil and pan, though."

Bob explains, "Our installation crew should have left a note on the thermostat to not turn the unit on while sanding floors or sheetrock. Oftentimes the sanders get hot and turn the system on anyway but pretend they didn't. That's not unusual. The secondary drain pan should have prevented the problem from happening."

Bob tells the owner about the secondary drain line float shut-off switch that could be added. The owner says, "Install it. The extra protection will be worth it."

By this time, Bob's crew from the shop has the den cleaned up and the owner is on the phone to a contractor to get an estimate on the ceiling repairs.

As they are driving away from the job, BTU Buddy says, "Service technicians often don't pay enough attention to drain line problems. A stopped-up drain can really cause a lot of damage. A system will normally drain about 3 pounds of water per hour per ton of air-conditioning. A pound of water is 1 pint. This was a 3-ton system, so it would drain about 9 pints per hour, times 24 hours of operation, which would be 216 pints, or 216 divided by 8 pints per gallon which equals 27 gallons. That will make a huge puddle. It would weigh 8.33 pounds per gallon times 27 gallons which equals 225 pounds. That's what brought the ceiling down."

Bob answers, "From now on, I'll take particular care to check condensate drain systems. It never occurred to me that they were that important or that they handled that much water."

# Outdoor Unit with Restricted Airflow

The dispatcher calls Bob and gives him a service call at a residence. The house is getting hot because the air-conditioning unit is not cooling even though it has been running. It's a 4-ton system.

The weather is oppressive when Bob arrives, 96°F. He pulls his truck to the back of the driveway to prevent blocking the homeowners' cars. Bob is familiar with this system. He goes to the front door and the wife answers. She invites him in to check out the system. The house is hot. The thermostat's thermometer registers 84°F. She tells Bob that the system had been working until it got really hot. He can hear the indoor fan running, but the air coming out of the registers is room temperature, not cold.

Bob goes out back to the outdoor unit and finds the problem quickly. The homeowners added an extension to the wooden deck in the back of the house and the unit is now under the deck. He hears the condenser fan running, but cannot hear the compressor. Bob goes into the house and turns the unit off to let it cool down. He then asks the owner when the deck was added on. She says that it was finished 3 days ago.

Bob asks her, "Did the workmen mention to you that they were covering the air conditioning unit with the deck?"

She responds, "Yes, but they said that it would be okay because there was plenty of room to work on the unit."

Bob then says, "I think the deck is the problem. There is only about two feet of clearance above the unit and the fan discharge is upward. Hot air must be hitting the deck and recirculating back into the unit, which would cause the compressor to overload and shut off," Figure 19-1. "I'll go to the unit and see what I can do."

She replies, "We're having a party here tomorrow night and we just have to have cooling. It's a very important business party."

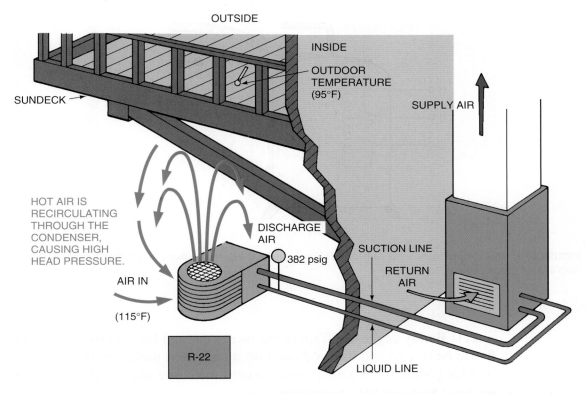

OUTSIDE

INSIDE

OUTDOOR
TEMPERATURE
(95°F)

SUNDECK

SUPPLY AIR

HOT AIR IS
RECIRCULATING
THROUGH THE
CONDENSER,
CAUSING HIGH
HEAD PRESSURE.

DISCHARGE
AIR

382 psig

SUCTION LINE

RETURN
AIR

AIR IN

(115°F)

R-22

LIQUID LINE

**FIGURE 19-1.** The hot air from the condenser discharge hits the deck and recirculates to the unit, causing it to overheat.

Bob goes down to the unit. Sure enough, the deck is about 18 inches above the fan discharge. This could never work.

BTU Buddy then appears and asks, "What's the problem, Bob?"

Bob explains what is going on. Either the deck or the unit must be moved. There's a party tomorrow night and the company doesn't have enough time to move the unit. "I have to have another plan."

BTU Buddy says, "Hook up a water hose and get the compressor cooled down while we work out the details."

Bob shuts off the electrical disconnect and locks it, then connects the water hose to cool the compressor, Figure 19-2.

BTU Buddy then says, "Call the shop and have them send out one of those large prop-type fans, and we can set it under here to move air over the condenser until we can work up a plan. Those fans move a lot of air and they're not very loud."

Bob makes the call and the fan is expected within half an hour. The compressor is cooling down.

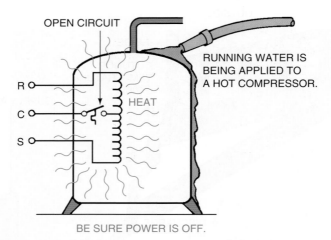

**FIGURE 19-2**. This compressor is being cooled quickly by opening the electrical disconnect to the unit and cooling it with a small trickle of water.

BTU Buddy suggests, "Talk to the owner and tell her that the condenser has to be moved and that you can't do it until next week, but that you can provide cooling with a temporary rig until then. She may want to have her husband look at what's happening."

Bob talks to her and she calls her husband who will come home in a few minutes.

BTU Buddy says, "Let's look at an alternative location for the condensing unit."

## RELOCATING THE CONDENSING UNIT

Bob and BTU Buddy walk around the back of the house close to the deck, looking for a good spot for the condenser. It looks like it should be moved to the end of the house and around the corner. BTU Buddy says, "That looks like a good place, but it's by the master bedroom and there are windows close by. I think you would get noise complaints if we moved it there."

The husband gets home about that time and joins the conversation. He suggests a particular spot that he likes. Bob looks up and sees that there is a drain right overhead and says, "This place is really not suitable because an excess of water could run into the unit," Figure 19-3.

The husband then asks, "Aren't they rainproof?"

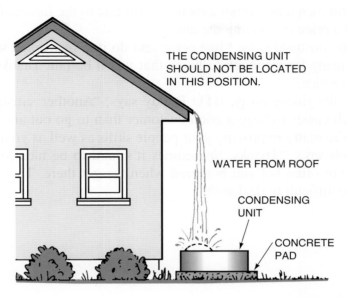

THE CONDENSING UNIT
SHOULD NOT BE LOCATED
IN THIS POSITION.

WATER FROM ROOF

CONDENSING
UNIT

CONCRETE
PAD

**FIGURE 19-3** Outdoor units must not be located where large amounts of water can pour into them. They can operate in the rain, but not large streams of water.

Bob says, "Yes, but there are times when large volumes of water will come down between these two roofs and pour into the unit. It isn't made to take that much water."

They continue looking around and decide to move the unit to the far end of the deck, next to the driveway. It's about 25 feet to the new location, but the piping and the electrical can be run under the house. It's actually closer to the electrical panel so the electrical run will be much closer.

The fan arrives and Bob removes the water from the compressor and replaces the compressor compartment cover. He goes into the house and sets the thermostat to cooling. He then turns on the large fan about five feet from the condensing unit. He turns the disconnect to the unit back on. The compressor and fan start. Bob feels the compressor suction line and notices that it is getting cool. The unit seems to have a charge of refrigerant in it. That looks good.

Bob asks the husband to come out to the unit so he can show him what's happening. Bob turns off the fan for a few minutes and the heat under the deck begins to build up. The husband says, "Wow, I'd have never believed it could get this hot this quickly."

Bob restarts the fan and the heat dissipates. The system will be okay as long as the fan is running.

Bob then tells the homeowners, "I'll talk to the foreman and have him get you a price on moving the unit."

The husband says, "Just come and do it. I've worked with you and your company long enough to know that you'll be fair. Thanks for the thoughtful service."

While riding away, BTU Buddy says, "Another satisfied customer. It's much easier to keep a good customer than to go out and find another one. You're really improving your people skills as well as your technical skills."

Bob says, "Thanks. Sometimes it's hard to be nice to customers, since they're often hot and bothered when we get there. That can make them a little difficult to deal with."

# Repairing a Restricted Liquid Line

The dispatcher calls Bob with a no cooling call. Actually, the unit had been cooling until last night, when the home began to get warm inside.

Bob arrives and talks to the homeowner about the events. She tells Bob that it was cool last night, but it began to get warmer in the early morning hours. The outside night temperature was 75°F, so some cooling was needed.

Bob goes to the outside unit and finds that it's running, but the suction line is frozen back to the compressor. He decides that there must be restricted airflow or that the system is low on refrigerant. He then goes into the house and turns the compressor off and the fan to "fan on" to thaw out the coil. This will take about 30 minutes, so he decides to check the rest of the system.

He determines that he'll check the filter first so he straps on his tool belt, takes a flashlight, and goes under the house to the air handler, which is a gas furnace. When he crawls to the furnace, he hears the indoor fan running and it sounds normal for a fan that's running against a coil that is frozen solid.

Bob shuts the unit off, removes the filter, and discovers that it looks good. It has been changed recently. So he replaces it and starts the fan again to help thaw the coil. There's not much to do now but wait until the coil thaws, so he goes in to the owner and explains what is happening. He tells her that it's going to take time to thaw the coil and so he's leaving to go to lunch in the meantime.

After lunch, he returns to the job and finds that the coil has thawed and the system is ready to start up. He fastens gauges to the unit with a cylinder of refrigerant so he'll be ready to add refrigerant to the system. About this time, BTU Buddy steps in and says, "Don't you think that you should do a little leak-checking on the system before adding refrigerant? You may just have to take it all out again to repair a leak."

Bob says, "Good idea; the system seems to have plenty of pressure. It isn't out of refrigerant; probably just low."

Bob goes to his truck for his electronic leak detector and then checks all of the field connections on the outdoor unit but can't find anything. He then crawls under the house and checks all of the connections at the coil. The piping at the coil is on the far side of the air handler, so he has to crawl around the duct to the piping side of the coil where the suction and liquid line and the condensate line are connected. He doesn't find any sign of a refrigerant leak on the piping connection or in the condensate line connection, so he comes back outside.

BTU Buddy asks, "What did you find?"

Bob explains, "I didn't find a leak, but I did find something a little strange. I had to dry the liquid line connection and the suction line connection. I can't figure out why the liquid line was wet. It wasn't under the suction line."

BTU Buddy says, "You may have discovered the problem, but I'm going to let you figure it out. Start the system and see what the pressures are."

Bob starts the compressor from the inside thermostat and comes back outside as the pressures begin to register.

After a few minutes, Bob says, "The suction pressure is 70 psig and this is R-410A refrigerant. That equates to a temperature of 15°F for the coil. This system is going to freeze again. It's definitely low on refrigerant."

BTU Buddy then says, "Before you add refrigerant, go under the house and check that liquid line again. I'll talk to you through the crawl space vent where the lines go through the foundation."

Bob goes under the house and finds that the liquid line is freezing cold. He goes to the vent and tells BTU Buddy what he has found.

BTU Buddy tells Bob, "Follow the cold line back to this wall and see if you find anything."

Bob starts looking and then says, "There's a kink in the line about three feet from the wall. The freezing starts there."

BTU Buddy tells Bob, "Come out and let's talk about it."

## A KINK IN THE LINE

When Bob gets out from under the house, he says that someone must have been working under the house and pushed the line up next to the floor joists, bending the line. The bend or kink is not really severe, but must be repaired, Figure 20-1.

**FIGURE 20-1.** This piece of bent tubing is similar to what Bob found under the house, causing a liquid line restriction.

BTU Buddy explains, "The kink in the line is a restriction, just like the metering device. So you now essentially have two metering devices in series; a lot of pressure drop. The reason the coil and suction line froze was that the low evaporator temperature made a block of ice out of the coil and it kept getting bigger until it became a solid block and the suction line froze.

"You have two choices: shut the system down and replace the bend with a coupling, or straighten the pipe out and use a flare block to round it out again. If you replace the bend with a coupling, you'll need to recover the refrigerant to make the repair."

Bob goes and talks to the owner and explains the options. He explains that the rounded-out pipe should last as long as the system and not be a problem as long as the pipe is not pushed around again.

The owner tells Bob that the cable repairman was under the house yesterday, connecting cable to that area, and that's likely when the line was bent. She tells him to try to round the pipe, which would be less expensive, as they're tight for money at the moment.

Bob shuts off the system to reduce the liquid line pressure, then gets his gloves, safety glasses, and flare block, pictured in Figure 20-2, and goes under the house. He gently straightens the pipe and all seems well. He then places the flare block around the pipe at the bent section and gently tightens the block down on the pipe. Little by little, it becomes round again with the pressure from the flare block. When he removes the block, the pipe looks pretty good, not perfect, but usable. He then fastens that length of pipe to the floor joists securely to protect it in the future. He fastens the rest of the liquid line to the joists until it approaches the coil where it must drop down to the coil.

Bob then goes outside and turns the unit on again. The suction pressure then settles down to 142.5 psig, which corresponds to a 50°F coil. It is hot in the house. The pressure will come down as the house begins to cool down.

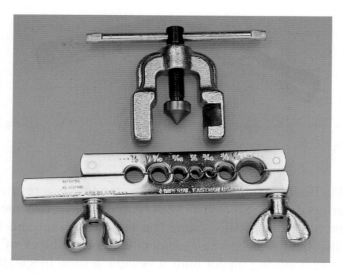

**FIGURE 20-2.** This flaring block can be used to make tubing round again. It must be used very carefully and, if the tubing is under pressure, be sure to wear goggles and gloves.

Bob writes up the report and goes in to the owner and says, "All looks well, I fastened that pipe up to protect it so this will not happen again."

She looks at the bill and says, "This is really fair for getting our system cooling again. Thank you."

Bob says, "It is not only just cooling, it is reliably cooling. It is going to last."

When driving away, BTU Buddy says, "That was a good job, at a reasonable price for the customer. All customers are looking for value and reliability in what they pay for. A job well done."

Bob says, "Thanks, it's good to see a customer that is happy."

# High-Efficiency Cooling in Mild Weather

Bob receives word from the dispatcher that a loyal customer has requested to speak to him in person regarding his system.

Bob stops in to see the customer, who explains that he is having a very important business event at his home on Saturday night. In order to be sure that his air-conditioners are operating correctly, he asks Bob to be in attendance to watch over the equipment during the event. The units are high-efficiency heat pumps, but they're about 10 years old. Bob is very familiar with this customer, who explains that the entire house must be comfortable, at about 72°F. There are two 5-ton heat pumps and one 3-ton heat pump that serve the party room where people will tend to gather. All of the air handlers and outdoor units are in remote places where Bob will not be noticed.

When Bob arrives at the event, he parks his truck out of sight in the back, and observes that the outdoor temperature is 76°F. It will get much cooler after dark; by 7:00 p.m. the temperature will be 60°F. He goes to the air handlers and places a thermometer in the supply plenum and return plenum of each unit, Figure 21-1. He decides to set the fans to "fan on" so he can monitor each unit during the event with a constant airflow. Everything is set up and he waits for the party to begin.

The equipment for this large home is:

- Unit 1: 5 tons serving the living room, kitchen, large den, dining room, two guest bedrooms, and entrance hall.
- Unit 2: 5 tons serving the upstairs with 4 bedrooms, a kids' playroom, and a den.
- Unit 3: 3 tons serving the large party room and a second kitchen used for events.

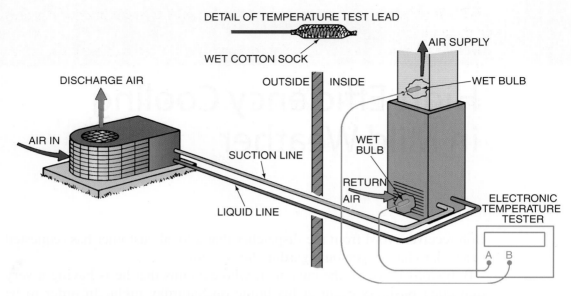

**FIGURE 21-1.** This illustration shows the correct placement of thermometers for this kind of test.

Thermometer readings before the event starts are:

- Unit 1: 73°F in and 56°F out; the unit is running.
- Unit 2: 72°F in and 72°F out; the unit is off.
- Unit 3: 73°F in and 56°F out; the unit is running.

Bob sits back with a soda and thinks that this will be some easy overtime for hanging out and watching equipment on a Saturday night. He doesn't know that he is about to earn his pay.

As people begin to arrive, the temperature outside begins to drop. Then the downstairs portion of the house is full of people and things begin to change. The readings go to:

- Unit 1: 74°F in and 60°F out; the room temperature is rising.
- Unit 2: 72°F in and 72°F out; this is fine.
- Unit 3: 74°F in and 60°F out; the room temperature is rising.

Units 1 and 3 are losing ground, it seems. Bob is getting concerned because the leaving air temperature is rising, as if the units are losing capacity. Bob is scratching his head and starts for his truck to get a set of gauges, when BTU Buddy appears and asks, "What's the problem, Bob?"

Bob shows him the readings and says, "The temperature is going up. I have to do something fast. They're depending on me and the equipment is losing ground."

BTU Buddy begins to ask questions. "What type of metering devices are on the units?" he says.

Bob answers, "Unit 1 is a thermostatic expansion valve (TXV), Unit 2 is a TXV, and Unit 3 is a fixed-bore orifice. What difference does it make?"

BTU Buddy then asks, "What is the outside temperature?"

Bob puts a temperature lead outside and it reads 60°F.

BTU Buddy says, "Let's take one unit at a time. Unit 3 is where the real party is going on. The temperature inside is 74°F, and the outside temperature is 60°F. Does something seem different to you about this?"

"The temperature outside is cooler than the temperature inside," Bob says.

BTU Buddy says, "Yes, that's true. What happens to condenser efficiency when the condenser is operating under cool conditions like this?"

Bob says, "It will become very efficient."

BTU Buddy then says, "Have you ever thought that it would become so efficient that the head pressure would drop low enough that it wouldn't push enough refrigerant through that orifice?"

"No, I would have never thought of that," Bob says.

"Feel the suction line and tell me what you think," says BTU Buddy.

Bob feels the suction line and says, "It's not cold. It feels like a low charge."

BTU Buddy then says, "Let's block some of the air going across that condenser and see what happens."

Bob gets a plastic garbage bag and wraps it partway around the condenser. BTU Buddy says, "Stand here for a few minutes until the air coming out of the condenser feels warm compared to your hand. Your hand temperature should be about 91°F. If the air is slightly warmer than your hand, you'll know that heat is really coming out of the condenser. The air wasn't warm at all when we first walked out here."

Bob asks, "Is there a chance that I might overload the system by doing this?"

BTU Buddy says, "Also feel the liquid line and let it get just warm to the touch, then you'll know that the head pressure is rising but isn't too high, as long as the liquid line is just warmer than your hand. Let's go in and look at the temperatures now. As you know, I don't like to use gauges unless it's necessary, because I've seen the charge altered using gauges and I've also seen too many technicians leave behind leaks by using gauges when it was not necessary. Gauges should only be used when needed."

He continues, "Now feel the suction line and describe it to me."

Bob grips the suction line and says, "It's cold, as it should be. It no longer feels like a low charge." Figure 21-2 depicts a similar evaporator.

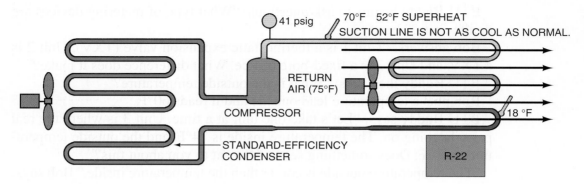

**FIGURE 21-2.** This evaporator is slightly starved of refrigerant. This could indicate a low charge, or as in this case, just a starved evaporator.

They go back inside and the Unit 1 readings are 74°F in and 55°F out.

Bob says, "Boy, what a difference that made. I still don't understand what happened. But before you answer, let's go do the same thing to Unit 3 and get the party room cooling down."

While they're putting plastic around the condenser to Unit 3, BTU Buddy says, "Unit 1 is cooling correctly at this time. It has a limited charge, meaning that during the cooling cycle, there should be a certain amount of refrigerant in the evaporator (the indoor coil), and a certain amount of refrigerant in the condenser (the outdoor coil). You do remember that from school, don't you? Even though this unit has a TXV, it does not have a large receiver to hold extra refrigerant like some of the older units did."

Bob says, "Yes, the unit charge is correct."

BTU Buddy says, "The charge is correct for a design day that is, for example, 75°F and 50% humidity for return air entering the evaporator, and 95°F for outside air entering the condenser. At that time, there will be a certain amount of refrigerant in the evaporator and the condenser. With the outdoor temperature at 60°F, the condenser will become overly efficient and liquid refrigerant will back up in the condenser. Where do you think it will come from?"

Bob says, "The only possible place would be the evaporator."

"Correct," says BTU Buddy. "Now if there is more refrigerant in the condenser and it's coming from the evaporator, the evaporator must be starved for refrigerant."

Bob then says, "I see what you mean. If it's starved, it will lose capacity. And this system is losing capacity just when it really needs it. The only way to get it working is to raise the head pressure and push that refrigerant through the metering device to the evaporator where it can do its work."

BTU Buddy says, "Now you have the picture. It was a simple fix, but if you hadn't been here to operate the equipment, the owner would have likely kept turning the thermostat down and eventually gone to bed later on without turning it back up. At the same time, the space temperature would have been uncomfortable. You earned your pay tonight."

Bob then says, "But you have to know what to do to operate equipment out of its design parameters. You made the difference. I'll never forget this night."

They go back inside to look at the readings, and Unit 1's air temperatures are 73°F in and 53°F out. Unit 3's temperatures are 73°F in and 53°F out.

Bob says, "This is really great. The units are gaining on the load now. You never did answer my question about the TXV versus the orifice expansion device, though."

BTU Buddy explains, "Both the orifice and the TXV metering device have a critical charge in a heat pump system. The TXV is much more forgiving than the orifice. The orifice will starve the coil much faster than the TXV, but neither one of them will operate well under the conditions that this job demanded."

Bob responds, "This has been a real learning experience and you've been a great help."

At the end of the night the owner comes out to talk to Bob and says, "Bob, you must have done a good job, because the conditions were just great in the house for the whole event. Thanks."

As Bob is riding away, BTU Buddy says, "Bob, that was a job well-done. You have a very satisfied customer."

# Frozen Evaporator on a 15-Ton Cooling Unit

Bob receives a call at about 9 a.m. that a restaurant's system is not cooling the dining room where a men's breakfast is being held. This is a new customer for his company.

When Bob arrives, he goes to the restaurant owner to talk to him and see what he can find out. The owner tells Bob that a banquet had been held in that same dining room last night and that the system had not cooled the room then either.

When Bob goes to the roof, the outside air is still cool from the night before. It had gone down to 55°F. The system is a split system with the evaporator and air-handling unit in a downstairs closet. Bob can hear the unit's fans and compressors running. The system has two compressors of 7-1/2 tons each. When he gets close to the unit, he notices that the suction line is frozen solid. The compressors are solid sheets of ice as well.

Bob goes down to the space temperature thermostat and turns the unit to "fan on" and turns the system to "off" to stop the compressors. The temperature setting for the thermostat is 55°F. He's pretty sure the problem is that they were trying to operate the unit below the design parameters, but he wants to confirm that. He is sure that the system froze because of a combination of the thermostat being set too low and the low outdoor ambient temperature. The problem is figuring out what to do about it to prevent it from happening again.

Bob explains to the owner that it will take about an hour to defrost the coils, and that he'll go to another job that's nearby during the defrost time. The owner thanks Bob for keeping him informed. He tells Bob that he has used other service companies before and that he was often annoyed that the technicians would just leave without an explanation.

When Bob returns, the system is thawed and ready to run. Since this is his first trip to this customer site, he decides to look the unit over carefully. The system is about 2 years old and looks to be in good shape. He checks the filters and they are clean. The owner had said that he changed them on a regular basis. Bob lubricates the evaporator fan motor and is ready to start the system. He's concerned that the system may have a low charge, so he decides to check it out. This system has two coils and two thermostatic expansion valves as metering devices.

Bob starts the system and goes to the roof to see how it's reacting. When he gets to the roof, the suction line is cold on both compressors. The system seems to be running normally for a morning that is 70°F. He's standing there thinking when BTU Buddy says, "What's the problem, Bob?"

Bob says, "This system froze solid last night because the thermostat was turned down to 55°F and the outdoor temperature was too cool to operate the system, even though it has thermostatic expansion valves. I am not sure where to start."

## POSSIBLE SOLUTIONS

BTU Buddy suggests, "You have three possible solutions, all of which I think would work in this case." His suggestions are as follows:

1. Put a lock box on the thermostat and tell the manager to only give one informed person the key. The turned-down thermostat problem often happens at night when a night shift person doesn't realize the consequences. Many people think that if you turn the thermostat to a lower position, the unit cools faster. They think the thermostat is like an accelerator on a car. They don't realize that it's like a light switch.
2. Install a "low-ambient" control to monitor the condenser fan speed, or cycle the condenser fans to maintain a minimum head pressure, Figure 22-1.
3. Install a freeze control on the suction lines to shut off the compressors when a freeze condition is sensed, Figure 22-2.

"These controls will protect the system and give proper control under low-ambient conditions. When an evaporator starts freezing, the situation is compounded by the ice on the coil and it then spirals out of control until the whole low-pressure side of the system is frozen solid," Figure 22-3.

"I would suggest that you contact the manufacturer for a low-ambient kit for this particular unit. One is probably available. The freeze control is an extra for the best protection."

**FIGURE 22-1.** This is a condenser fan cycle control. It is one of many ways that the head pressure is maintained during low-ambient conditions.

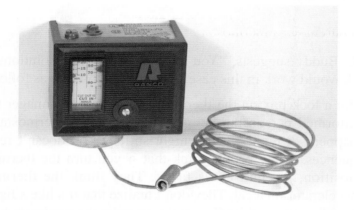

**FIGURE 22-2.** The bulb on the freeze control can be mounted on the suction line of the compressor and will shut the compressor off when freezing conditions occur and turn it on again when the line warms up.

Bob goes to the owner and explains what's been happening. The owner asks him for a quote to make the control system functional under these conditions. Bob goes to his truck to figure it all out, and comes back to him with a quote. The owner gives him permission to proceed.

Bob is able to get a low-ambient control kit for the unit, which straps onto the liquid line and measures the temperature, and cycles the condenser fan at the appropriate time. This is much better than using a low-ambient

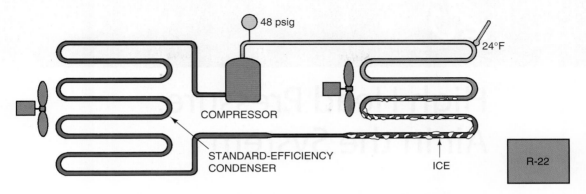

**FIGURE 22-3.** This coil is beginning to freeze and accumulate ice, which will restrict the airflow and cause more ice to form. From there it will only get worse until the low-side lines are frozen and even the compressor will freeze over with ice.

kit that uses pressure as the sensor, as he won't have to create a pressure source on the system to operate the control and possibly create a leak. He also purchases a thermostat cover that has a lock. The cover has good ventilation so the thermostat will have good airflow over the sensor. He also purchases two freeze controls, one for each compressor.

Bob installs all of the controls and starts the system, and it operates correctly. He goes to the owner and says, "Your system is now protected from the conditions we encountered on this call. If your employees turn the thermostat down again, the system will be protected by the freeze controls. For the thermostat cover, please keep one key for management and one in a safe place in case the other is lost."

The owner says, "Thanks for giving my system a good examination. It's nice to know that it's in the hands of a professional."

As they are riding away from the job, BTU Buddy says, "You did a good job of taking care of that customer. I think that he'll request you for the next job."

Bob then says, "It really feels good to leave a job knowing that you did it right and that the customer is satisfied."

# High Head Pressure, Air in the System

Bob receives a call from the dispatcher that a new customer has called. It's 95°F outside and the homeowner's air-conditioner is not working. The fan outside is on and the fan inside is on, but no cool air is coming out.

When Bob arrives, the customer says, "The unit stopped last week and a service man from another company came out and made the repair. I didn't like the service technician. He was very unprofessional. He made a quick repair and left."

Bob asks, "May I see the service ticket?"

The customer shows him the ticket describing the work. The ticket reads, "Suction line vibrated and had a large leak. Soldered the leak and charged the system."

Bob goes to the unit to see if he can determine where the leak was. The condenser fan is running, but the compressor is not. He removes the compressor access door and finds a patched area on the suction line where the cabinet had rubbed a hole in the line. It seems to be well-patched with 15% silver solder. The line had been moved to where it would not touch the cabinet anymore. About this time the compressor starts and Bob can hear it begin to work. He reaches over and feels the suction line, which is cold. It seems a little too cold, as though it has liquid refrigerant in it. The unit continues to run. He touches the liquid line. It's very hot.

Bob is standing there scratching his head when BTU Buddy appears. "What seems to be the problem, Bob?"

"Things aren't adding up. The unit seems to have too much refrigerant, because the suction line is cold, but the liquid line is hot. It seems that if this were an overcharge, the liquid line would be cool from extra subcooling."

BTU Buddy then says, "I think it's time to put your gauges on. Just shut the system down until you're ready."

About that time the compressor shuts off, but the condenser fan keeps running.

Bob says, "Well, I guess that solved that; the compressor shut itself off."

BTU Buddy then says, "Turn off the power for a moment and let's try something."

Bob turns off the power. "Now what?"

BTU Buddy says, "Remove the common lead to the compressor and tape it with electrical tape, then turn on the condenser power again. We want the fan to run, but without the compressor."

Bob does as BTU Buddy suggests and asks, "What now?"

"Put your gauges on the high and low side," BTU Buddy says.

Bob fastens the gauges.

The condenser fan is running and Bob is observing the gauges when BTU Buddy asks, "Do you see anything strange or different about the gauge readings?"

Bob says, "No, what do you mean?"

BTU Buddy says, "What do you think the readings should be? This is an R-22 system. Think for a minute; the outdoor unit has liquid refrigerant in it and the fan is running, passing 95°F air over the coil. What could that pressure be? The indoor fan is running and passing 85°F air over the coil. What could the pressure be?"

Figure 23-1 is an illustration of a standard-efficiency system and a high-efficiency system operating under design conditions.

Bob thinks for a minute and says, "Well, the unit has a capillary tube metering device that should equalize between the high and the low side. I would think that most of the liquid would eventually move to the indoor coil where it's the coolest. The high-side pressure should not exceed the level corresponding to 95°F, which would be 182 psig. The lowest pressure reading should correspond to the air temperature passing over the indoor coil, which is 85°F or 155.7 psig. Hey, would you look at that—the high-side pressure is 193 psig."

BTU Buddy asks, "What could that mean?"

Bob responds, "Either the unit has some air in it or there's another refrigerant in it."

BTU Buddy then says, "Read the last repair ticket carefully."

Bob says, "He mentioned a low-side leak, but didn't mention that he had evacuated the system. I'll bet this system has air in it. It would have run

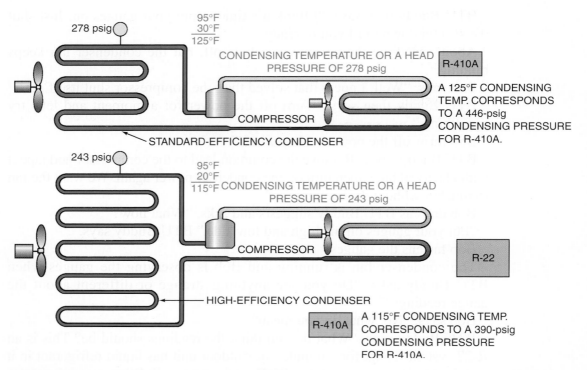

**FIGURE 23-1.** This figure shows both a standard-efficiency system and a high-efficiency system operating at design conditions.

when the charge became low, because it doesn't have a low-pressure shut off control. The only thing that would have shut the unit down would be the motor temperature thermostat."

"Good observation," remarks BTU Buddy. "Now what?"

Bob answers, "This system needs the refrigerant recovered, a drier added to the liquid line, and then it needs to be evacuated and charged."

"Correct," BTU Buddy says.

Bob goes to the homeowner and explains what needs to be done, without bad-mouthing the other service technician. She says to Bob, "Repair it to the best of standards. You don't have to say anything about the other technician. I already had him figured out."

After the repairs are made and the system is running, Bob says to BTU Buddy, "That was a good find. When we arrived, the compressor was running at part capacity and shutting off from time to time because of compressor internal overload. Then it would start up and run for short periods of time. I still don't fully understand how the air acted in the system."

BTU Buddy explains, "There are a few guidelines that every technician should remember. Anytime there is a possibility that air could have entered a system, assume that it did and take the maximum recommended precautions. Air contains moisture and oxygen as well as nitrogen. The nitrogen makes up about 80% of the air. The oxygen will cause mild acid to form in the system and eventually cause internal electroplating of a soft metal to a hard metal. All of the ingredients are there for electroplating: electricity, dissimilar metals, and acid. This would normally be copper to steel and would occur on the bearing surfaces, causing them to become larger. Then the bearings become tight and begin to wear." Figure 23-2 shows the crankshaft and rods where this electroplating occurs.

"A more immediate response is due to the oxygen gas and the nitrogen gas which will not condense at the temperature and pressure at which the system operates. They move quickly to the condenser where they will not condense. The system has a liquid seal between the low- and high-pressure side (the expansion device), so the non-condensing gases stay in the top of the condenser, taking up condensing space. This effectively makes the condenser undersized, and the head pressure rises, causing the compressor to shut down on motor temperature thermostat."

Bob says, "So what you are telling me is that an immediate reaction caused high head pressure and a safety shut-down, and if there were just a little air in the system, long-term damage would occur due to electroplating. I guess we were lucky that the air in the system was great enough to cause us to make a repair when we did."

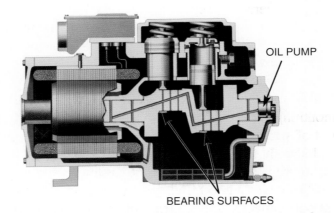

**FIGURE 23-2.** This drawing is of a compressor that shows the most likely areas for electroplating to occur. When the bearings become larger due to plating, they will begin to bind and wear. *(Courtesy Trane Company.)*

BTU Buddy says, "Yes, in the long run, the system could have suffered with only a small amount of air. Well, the system is running now with the correct charge of only refrigerant and oil, it's clean and dry, and should last for years without problems."

Bob goes to talk to the customer before leaving and says, "Your system is in good shape and should provide you with many years of good service."

She says, "Thank you for being very professional about this repair. We'll be using your company from now on."

As they are driving away, BTU Buddy says, "Another customer that likes your services. They keep adding up. Your company is noticing the business that you're bringing to them."

Bob says, "Thanks for the suggestions and help."

# Stopped-Up Condensate Line

Bob receives a call from the dispatcher that the basement in a large home is flooded, and the owner thinks it has something to do with the air-conditioner. When Bob pulls into the driveway, the homeowner meets him at his truck and says, "There is a lot of water in the basement and it seems to be coming from the air handler that's located in a closet in the basement."

Bob has never done work for this customer, so he asks the owner to take him to this closet where he thinks the water is coming from.

When they enter the basement, there is about 1/4 inch of water on the floor around the air handler and in the adjoining room. The carpet is soaked. Bob tells the owner that the quicker he gets the water up, the better it will be. He recommends a carpet-cleaning company with a large vacuum system to extract the water from the carpet.

The owner goes to contact the carpet-cleaning company, and Bob starts to search for the problem. What he finds confuses him. The condensate drain runs into the concrete floor in plastic pipe, but he doesn't know where it terminates. He goes outside to look for where it drains into the yard, but can't find any sign of the drain. BTU Buddy appears and says, "It looks like there has been a lot of landscaping here recently. There are raised flowerbeds all around the house. That pipe could terminate anywhere. Let's ask the owner if he knows where."

Bob asks the owner, "Have you ever noticed any water draining in the yard that could have come from the air conditioner?"

The owner says, "No, the yard has always been dry."

BTU Buddy then says, "I'll bet the drain line went to a dry well that is supposed to absorb the condensate water. The dry well might be saturated, or the landscape help may have dug it up and planted a tree in the hole," Figure 24-1.

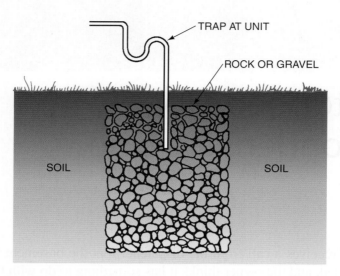

**FIGURE 24-1.** This is a dry well. It is a hole filled with rock or gravel. The soil must be able to absorb the water or the well will eventually fill up.

Bob then asks, "Wouldn't they have said something to someone first?"

BTU Buddy responds, "Maybe not. They could have just assumed it was their job to grade the earth and plant."

"Now what?" Bob asks.

BTU Buddy explains, "There's no good way to drain the condensate out by gravity. There should have been a floor drain installed here; instead, the water was carried out to the yard by gravity. Now it's uphill from the drain on the evaporator coil to any place else. We need a condensate pump to pump the water uphill to a termination point. Let's look around and see if there's a laundry room down here," Figure 24-2.

Bob looks and finds a laundry room down the hall. BTU Buddy then suggests that he get the owner involved in the decision.

Bob explains the situation to the owner and shows him a picture of a condensate pump that has a float that starts the pump when the small reservoir is filled with water. He also explains that the pump has another feature that will shut the air conditioner off in case the primary float doesn't function. This would prevent the flooding problem from recurring. He also shows the owner how he would route the small plastic pipe from the condensate pump to the laundry room, where it would terminate in the drainpipe for the washing machine.

The owner gives Bob permission to proceed. He then asks how much water a unit like this generates. Bob does some calculations and says,

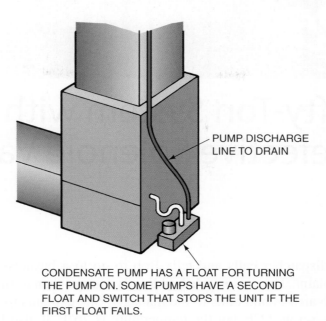

PUMP DISCHARGE
LINE TO DRAIN

CONDENSATE PUMP HAS A FLOAT FOR TURNING
THE PUMP ON. SOME PUMPS HAVE A SECOND
FLOAT AND SWITCH THAT STOPS THE UNIT IF THE
FIRST FLOAT FAILS.

**FIGURE 24-2.** This condensate pump may be used to pump small amounts of water uphill to a drain location. It is great for basement locations where there is no drain.

"A unit would normally generate about 3 pints per hour per ton of air conditioning. This is a 2-ton unit, so that would be about 6 pints per running hour. The weather is very hot, so I would suspect the unit is running 20 out of 24 hours. That is $20 \times 6 = 120$ pints per 24 hours, which is 15 gallons (120 pints ÷ 8 pints per gallon = 15 gallons). Fifteen gallons is a lot of water. This must have been stopped-up for a few days."

The owner says, "We often go several days without coming down here, so that explains why there was so much water."

"I'm going to get on with installing the pump," Bob says.

Before he leaves the job, Bob goes to the owner one more time and asks him to come and see the pump where Bob explains the installation.

The owner says, "Thanks for taking care of the job for me. The carpet crew is about through. They've sucked up all of the water and are setting up fans to further dry the carpet. They suggested that we keep the air-conditioning running to help remove the moisture."

As they drive away, BTU Buddy says, "That job turned out well even though it started with so many problems."

"Thanks for the help," Bob says. "Another satisfied customer makes it all worth it."

# Fifty-Ton System with a Defective Solenoid Valve

The dispatcher calls and tells Bob to go to a business that has a cooling complaint. The customer's building is warmer than normal.

Bob arrives and talks to the manager. She explains to Bob, "The thermostat is set at 72°F, but the temperature is 76°F. It also feels very humid in the building."

Bob notes that it is 96°F outside, but the system should be doing a better job than this.

The manager walks him down the hall and shows him the equipment room. She says, "The rest of the equipment is on the roof."

She leaves Bob to investigate the problem. He notices right away that there are two coils in the air handler because there are two thermostatic expansion valves (TXVs). He also notices that there are two solenoid valves, one just before each expansion valve. Bob has never worked on a system like this, so he's looking it over carefully. He notices a ladder and a roof hatch to get onto the roof, so he decides to go up there to see if he notices anything unusual. The compressor is running, so he removes the panel to it and feels around. The compressor seems to be running well and feels normal. He's standing there scratching his head when BTU Buddy asks, "Got a problem?"

Bob explains, "Everything seems normal, but the conditioned space is too warm and humid. The compressor is running and looks normal. I'm going to put gauges on the compressor."

Bob fastens his gauges to the compressor. The suction pressure is 74 psig and the head pressure is 185 psig. Bob then says, "The suction pressure is a little high and the head pressure is a little low. This looks like a bad compressor to me."

BTU Buddy tells Bob, "See what the amperage is on the compressor."

Bob checks the amperage and reports it to be 55 amps.

BTU Buddy then asks, "What is the full-load amperage on the compressor nameplate?"

Bob looks and says, "It's 100 amps full-load. This verifies that the compressor is not doing its job. Maybe some of the cylinders are no good."

BTU Buddy then says, "You're not familiar with a system like this. Let's go to the equipment room and do some checking."

## EXPLAINING CAPACITY CONTROL

When they move downstairs, BTU Buddy begins an explanation. "This system has capacity control. The total pumping capacity of the compressor and the rest of the system is 50 tons; however, it is capable of operating at several different capacities when needed. For example, it's the middle of the day and the load on the conditioned space is higher than normal. The compressor has 8 cylinders; 50 tons divided by 8 equals 6.25 tons per cylinder. The manufacturer has a method of unloading the cylinders so the system can operate at reduced capacity at reduced loads. For example, in the morning, the unit may only have a capacity of 25 tons when it starts up. Suppose the compressor could be made to run at 25 tons by operating on four cylinders ($6.25 \times 4 = 25$ tons)?"

Bob says, "That really sounds good. How can the compressor unload on demand?"

"Using a two-cylinder compressor as an example, what would happen if we could hold open one of the suction valves on one of the cylinders?" BTU Buddy asks.

Bob says, "Gosh, I don't know."

BTU Buddy says, "When the piston goes down on the suction stroke, what would happen?"

"The cylinder would fill with suction gas," Bob answers.

"Good," says BTU Buddy. "Now with the suction valve held open, when the piston comes back up for the compression stroke, it will push that suction gas right back out into the suction line. No work has been accomplished except overcoming friction." Figure 25-1 illustrates this.

Bob says, "With no work accomplished, the current draw or amperage would go down."

BTU Buddy says, "That's correct. Also, it's worth noting that the cool suction gas drawn into the cylinder will keep the cylinder cool and lubricate it, so it can run for long periods of time under these conditions.

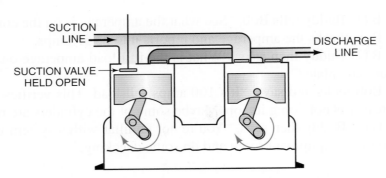

**FIGURE 25-1.** The suction valve on the left can be held open. When the piston rises in the compression stroke, the gas inside the cylinder is pumped back into the suction side. The cylinder is cooled and lubricated with this action.

"Now, let's talk about this system. When we want the system to run at reduced capacity, we merely need to shut off the refrigerant flow to one of the coils to reduce the flow capacity by half. The compressor has what is called suction pressure unloaders, meaning they are sensitive to suction pressure. When the suction pressure is reduced, the compressor will automatically unload to maintain the approximate suction pressure."

Bob reaches over and touches the solenoid valves on each circuit and says, "One valve is hot and the other is room temperature. This valve must not be working."

BTU Buddy explains, "Before you change the valve, check to make sure there is voltage to it. This could also be a thermostat problem."

Bob checks the voltage and it's 115 volts, the control voltage for this system. "There is voltage," he says, "The solenoid coil must be defective."

Bob goes to the supply house and gets a new coil for the valve. He shuts off the power and changes the coil, Figure 25-2.

Bob then turns the power back on and starts the system. He goes to the roof and checks the amperage on the compressor, finding that it's drawing full-load amperage. He turns to BTU Buddy and says, "Thanks for walking me through that one. I don't know when I would have figured it out on my own."

BTU Buddy then notes, "The reason that the humidity seemed so high in the office space is that the bottom coil is the one that failed. The top coil was working, taking moisture out of the air. But that moisture was running down over the bottom coil with the air flowing over it, causing most of the moisture to evaporate back into the conditioned space. In this type of system, the lower evaporator is the first stage—the first on and last off—and the upper

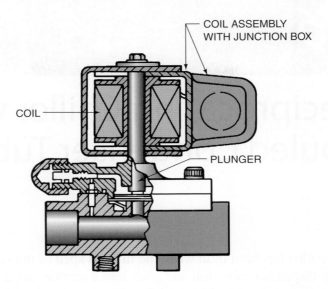

**FIGURE 25-2.** This is a typical solenoid valve that may be used to control liquid refrigerant flow.

evaporator is the second stage—last on and first off—on the thermostat. Always remember that, and don't get them reversed or the system will re-evaporate moisture into the conditioned space if the bottom evaporator shuts off and the top evaporator is still operating."

Bob says, "Well, another lesson. Thanks."

As they leave the job, BTU Buddy says, "The manager was really pleased that you got to the bottom of that call in good time and that it didn't cost too much. I think you have another satisfied customer."

# Reciprocating Chiller with Fouled Condenser Tubes

The weather has been mild when the first hot spell of the year suddenly moves in. The dispatcher calls Bob and gives him a service call at a motel. The motel has been operating its 100-ton chiller for about a month, but the weather has been mild. It's now the first 90°F day, and the chiller shuts down.

Bob arrives at the motel and the manager shows him to the equipment room where the chiller is located. It's not running. Bob begins to look around and he opens the control panel on the chiller to find that the manual reset high-pressure control button is out, indicating that high-pressure control has shut the chiller down, Figure 26-1. Bob reaches to reset the high-pressure control when BTU Buddy appears and says, "Why don't you check around a little bit before you reset the control. It would also be a good idea to fasten your gauges to the compressor so you can monitor the pressures when it starts up."

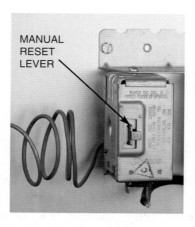

MANUAL
RESET
LEVER

**FIGURE 26-1.** A high-pressure control that is manual reset.

Bob says, "Well, that's just like me; just push a button without thinking. You're right. I should think more before I act."

Bob installs gauges on the compressor on the high and low side gauge ports. Then BTU Buddy says, "It's a good idea to check the water tower before you reset the control, to see how dirty it is and to see if there is proper water flow over the tower."

Bob goes to the tower and notices that there is plenty of water flowing over the tower, but the water is dirty and there's a lot of green algae scum in the tower basin. He tells BTU Buddy, "There's no sense in trying to start this chiller until the tower is cleaned up."

Bob calls the office and asks for a helper to come out and help him clean the tower. The helper arrives with rubber boots, face masks, and chemicals, and they start cleaning the tower and the strainer to the pump as well. They must use a lot of caution because the dirt in the tower can be unhealthy. After 2 hours they're finished cleaning the tower and basin and Bob refills them and starts recirculating the fresh water. Now they're ready to start the chiller.

Bob resets the high-pressure control and the chiller starts. After running for about 15 minutes, the head pressure begins to rise. The water from the cooling tower returns to the condenser at 83°F, and the water from the condenser back to the tower is 88°F, Figure 26-2. Bob scratches his head. BTU Buddy asks, "What's the matter? What do you see?"

Bob says, "I thought this chiller should add 10°F to the water and it's only adding 5°F. The chiller is operating at full load; the head pressure is 229 psig (R-22) and climbing. I'm not sure what I see."

BTU Buddy says, "Feel the liquid line going to the expansion valve and tell me what you feel."

Bob takes the line in his hand and says, "It's really hot, with 83°F entering water. You would think it would be cool."

BTU Buddy says, "It's obvious that the cooling tower water is not taking the heat out of the refrigerant. What could that mean?"

"I get it," Bob says. "There is plenty of capacity to remove the heat, but it's not being removed. The condenser tubes must be dirty, like the tower basin was."

"Correct," says BTU Buddy. "The only thing to do is to shut it down and clean the tubes."

Bob shuts the chiller down and removes the head from the condenser. The tubes are really dirty. He calls the office again, and they send out a helper with a tube-cleaning machine, shown in Figure 26-3, that turns a nylon brush and injects water at the same time. They clean the tubes and

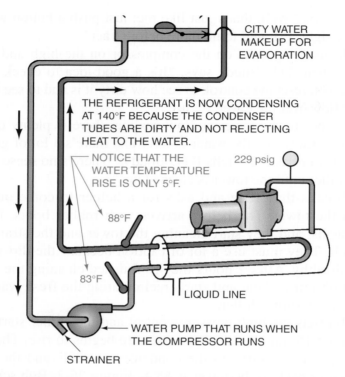

CITY WATER
MAKEUP FOR
EVAPORATION

THE REFRIGERANT IS NOW CONDENSING
AT 140°F BECAUSE THE CONDENSER
TUBES ARE DIRTY AND NOT REJECTING
HEAT TO THE WATER.

NOTICE THAT THE
WATER TEMPERATURE
RISE IS ONLY 5°F.

229 psig

88°F

83°F

LIQUID LINE

WATER PUMP THAT RUNS WHEN
THE COMPRESSOR RUNS

STRAINER

**FIGURE 26-2.** The cooling tower water is not taking the heat out of the condenser because the tubes are dirty.

**FIGURE 26-3.** This technician is brushing the tubes in a condenser with a machine that rotates a nylon brush while flushing the tube with water. *(Courtesy Goodway Tools Corporation.)*

flush out the condenser shell with fresh water, and then replace the head on the condenser. It's ready to start up again.

Bob starts the chiller while watching the gauges. The entering water to the cooling tower quickly rises to 85°F and the leaving water rises to 95°F. The chiller is operating well, as the outside temperature is now 95°F. Figure 26-4 shows a cooling tower and condenser.

"Well," Bob says, "every job seems to include something unexpected. You never get bored with all of the surprises."

BTU Buddy says, "You went about this with a systematic approach and were able to find the problem. Now it's time to recommend to the owner that they get a water treatment program going, as well as a service contract to keep the machine operating at peak efficiency. This machine has been costing them extra money operating at high head pressures for such a long time. They've been paying extra for poor efficiency."

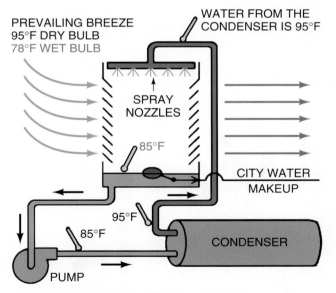

**FIGURE 26-4.** This cooling tower and condenser are working together to remove the heat from the system. Notice that the tower temperature difference is 10°F.

Bob gives the motel management a proposal, and they decide that it's a good idea. Bob has sold a service contract that is valuable to his company, and is a good value to the customer.

As they drive away, BTU Buddy says, "Well, Bob, your service call was a good value to the customer, and you've added business to your company volume. It's hard to beat that for a good day's work. Now everyone is happy. Every technician should try for this kind of day."

Bob says, "Thanks for the advice along the way."

# Topping Off the Charge for a Fixed-Bore Metering Device System

The dispatcher calls Bob and asks him to go to a new customer who is reporting a system that's not cooling up to capacity.

Bob arrives at the job and looks over the system. The thermostat is set at 72°F and the house temperature is 78°F. The unit has been running constantly during the day, but not cooling the house to a comfortable level.

Bob goes to the outdoor unit and feels the suction line entering the unit. It's cool, but not cold like it should be. He figures he can add a little refrigerant and be on to the next job, so he goes to his truck to get his gauges and a cylinder of R-22.

BTU Buddy appears and begins to question him. "What are you going to do?"

Bob replies, "Just add a little refrigerant. It's only a little low."

BTU Buddy says, "I'd suggest that before you remove the caps on the service ports, you check them for a leak. It's possible that they were left seeping refrigerant and that may be the source of the leak. There seems to be enough refrigerant in the system to leak-check them."

Bob turns the unit off to let the pressure equalize so there will be enough pressure on the low-pressure service port to leak-check that also. Then he goes to his truck for his electronic leak detector and gets set up to do the leak check.

Sure enough, the suction service port was leaking even with the cap on it. He removes the cap and uses his Schrader valve tool to tighten the valve stem core, Figure 27-1. The leak stops.

Bob says, "Well, I guess the last technician accidentally overlooked this leak. If I had just put the gauges on the system and charged it, we may

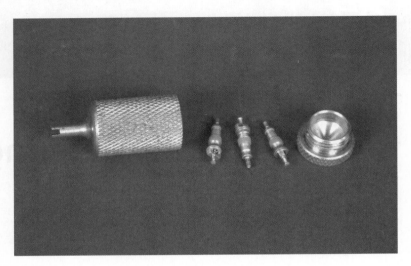

**FIGURE 27-1.** This service tool can be used for changing or tightening Schrader valve service stems. They are much like tire stems, except for the material they are made of. Notice this tool has a place for extra valve stems.

never have known about it. This system is only a little short of refrigerant, so that should be the only leak."

BTU Buddy responds, "It probably was the only leak. The other valve stem looked fine. Now, let's just add enough refrigerant to get the system up to peak performance. How do you propose to do that?"

Bob says, "One way would be to just add refrigerant until the suction line is sweating."

BTU Buddy then says, "Let's give this customer their money's worth and do the job just like you would want the job done at your house, when you are paying for the service call and the power bill. It's 85°F outside temperature. Let's get set up to charge the system as if it were a 95°F day, so when it does get up to 95°F, the charge will be correct and the system will perform at maximum efficiency."

Bob asks, "Do you mean the superheat method? This is a system with an orifice for a metering device."

"Yes," answers BTU Buddy. "The only way to add refrigerant is to check the superheat at the outdoor unit and build up the head pressure as though it were a 95°F day. What should the head pressure be for a 95°F day for this system?"

Bob looks at the unit and says, "It's obvious that the unit has some age on it. My pressure temperature card shows that for a 95°F day and a 30°F allowance for a standard efficiency unit, the head pressure should be about 278 psig."

BTU Buddy says, "Good math; 95°F outside temperature plus 30°F condensing approach temperature would mean that the refrigerant should be condensing at about 125°F, which corresponds to 278 psig for R-22."

Bob then fastens a temperature lead to the suction line and fastens his gauges to the gauge ports. The gauges show a head pressure of 225 psig (a condensing temperature of 110°F), a suction pressure of 55 psig (a boiling temperature of 30°F), and a suction line temperature of 60°F.

"The unit has a 30°F superheat," he says. "The evaporator is starved for refrigerant."

BTU Buddy agrees and says, "With a low head pressure and a low suction pressure, the system must be low on refrigerant. Let's add some, and block some of the condenser air to get the head pressure to rise."

Bob uses a piece of metal from his truck on the fan outlet to block the condenser airflow, and adds a small amount of refrigerant. The suction pressure begins to rise, as does the head pressure, When the head pressure reaches 275 psig, Bob begins to uncover the condenser fan outlet and maintains the 275 psig head pressure, Figure 27-2. The superheat is now down to 20°F with a suction pressure of 65 psig. The system is beginning to work now; the suction line is beginning to sweat.

BTU Buddy says, "Let's let the system run like this for a while until it settles down. The superheat isn't staying steady."

After the system runs for about 15 minutes, the superheat is showing 20°F. The system still needs some refrigerant.

BTU Buddy explains to Bob, "We're taking the superheat at the outdoor unit. We're really shooting for a superheat of 8°F to 12°F at the air handler where the evaporator is, which is about 25 feet from the outdoor unit. There will be some pressure drop in the suction line and some heat will be added to the refrigerant before it reaches the outdoor unit, so we need to compensate. With a suction line up to 30 feet we'll charge the unit for 10°F to 15°F. If the line set is longer than 30 feet, we would charge for a superheat of 15°F to 18°F."

They add a small amount of refrigerant again and after about 15 minutes, the superheat settles down to 13°F, Figure 27-2. The actual boiling temperature at the evaporator should be about 40°F, which is design temperature.

Bob says, "Well, there's a lot to getting the job done right. This homeowner will now have an efficient system."

BTU Buddy says, "The last thing that you should do is check the gauge ports for leaks. Then if the system loses refrigerant again, you'll know that the leak is somewhere else. It seems like technicians through the years have become careless about refrigerant leaks. Refrigerant was inexpensive for

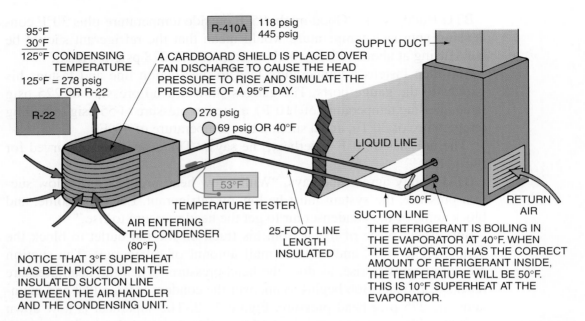

95°F
30°F
125°F CONDENSING
TEMPERATURE
125°F = 278 psig
FOR R-22

R-22

R-410A  118 psig
        445 psig

SUPPLY DUCT →

A CARDBOARD SHIELD IS PLACED OVER
FAN DISCHARGE TO CAUSE THE HEAD
PRESSURE TO RISE AND SIMULATE THE
PRESSURE OF A 95°F DAY.

278 psig
69 psig OR 40°F

LIQUID LINE

53°F

TEMPERATURE TESTER

AIR ENTERING
THE CONDENSER
(80°F)

NOTICE THAT 3°F SUPERHEAT
HAS BEEN PICKED UP IN THE
INSULATED SUCTION LINE
BETWEEN THE AIR HANDLER
AND THE CONDENSING UNIT.

25-FOOT LINE
LENGTH
INSULATED

50°F
SUCTION LINE

RETURN
AIR

THE REFRIGERANT IS BOILING IN
THE EVAPORATOR AT 40°F. WHEN
THE EVAPORATOR HAS THE CORRECT
AMOUNT OF REFRIGERANT INSIDE,
THE TEMPERATURE WILL BE 50°F.
THIS IS 10°F SUPERHEAT AT THE
EVAPORATOR.

**FIGURE 27-2.** This system is being charged on an 80°F day by blocking some of the air across the condenser to get the head pressure up to simulate a 95° day.

many years and technicians just got into the habit of adding it instead of finding leaks."

"The way the laws are today," Bob says. "The technician that can locate leaks and then repair them is a good technician."

BTU Buddy says, "On the next service call, we'll top off the charge for a high-efficiency system that uses a thermostatic expansion valve (TXV). This procedure is a little different because the TXV controls superheat and cannot be used to charge the system."

# Topping Off the Charge for a TXV System

Bob and BTU Buddy have met on another job involving a system that has a low charge. The season has gotten hot, and units that are not performing are beginning to show up. This unit has a thermostatic expansion valve (TXV). The charge is somewhat low and it's not cooling correctly. Bob and BTU Buddy are standing at the outdoor unit talking, when BTU Buddy asks Bob, "Just from looking and feeling around on the unit, what do you think?"

Bob says, "The suction line's not very cool; the compressor is very warm. I would say that the evaporator is starved for refrigerant. I'd suspect a low charge."

BTU Buddy then says, "Several things can cause the evaporator to be starved. The expansion valve could be out of adjustment. Tell me what the liquid line feels like."

Bob holds the liquid line in his hand and says, "It's hot. Shouldn't it be warm instead of hot?"

BTU Buddy explains, "The reason it's hot is that it has no subcooling. There is hot vapor from the condenser circulating in the liquid line. That's a sure sign of low charge. We're at the outdoor unit, touching the liquid line. If there were a restriction in the liquid line and it were starving the TXV for refrigerant, the liquid line would be cold just after the restriction. If there were a restriction in the liquid line, up next to the TXV, the liquid line leading to it would be warm. There would be a lot of subcooling in the liquid line because refrigerant would be backed up in the condenser."

He then adds, "Let's go in to the air handler and look for another clue."

When they reach the air handler, which is in the garage, BTU Buddy says, "Listen to the expansion valve and tell me what you hear."

"It's making a hissing sound," Bob says.

BTU Buddy says, "Yes, there's vapor passing through the valve—it doesn't sound like pure liquid. This tells us that the starved evaporator is the result of not enough liquid in the liquid line."

"So the hot liquid line is telling us that the refrigerant charge is low?" Bob asks. "And the hissing TXV is telling us that the starving for refrigerant is coming from the liquid line?"

BTU Buddy says, "Yes, using your hands and your ears to check out a unit is very effective. You can verify what your hands and ears tell you with your instruments. Let's see if we can find the leak. Turn the unit off and let the pressures equalize from the high-pressure side to the low-pressure side. If the expansion valve prevents them from equalizing, just use your gauge manifold to let them equalize. We need to have good pressure on the low-pressure side of the system in order to properly leak-check the unit."

Bob turns off the unit and performs a leak check of the service valve connections. He finds no leak, and so continues on to fasten the gauges to the system.

He fastens his gauge manifold to the unit, then purges the lines and connects the center line to a cylinder of R-22, the same type of refrigerant in the system. He shuts the system off and the pressures don't equalize, so he opens the gauge manifold valves and lets the pressure move from the high-pressure side to the low-pressure side. The equalized pressure on the system becomes 133 psig.

Bob says, "That's enough pressure to find a leak. The leak detector is warmed up so I'm going to start here at the outdoor unit."

The suction line is dry, so there's no need to wipe it off before leak-checking. He doesn't discover a leak at the outdoor unit, so he moves to the indoor unit. When he puts the leak detector probe into the area around the suction line, the detector shows a leak. He removes the cover from the side of the coil and finds that a hole has rubbed through the suction line.

Bob reports to the homeowner that the refrigerant will have to be recovered and the leak repaired. He then sets up his recovery machine, shown in Figure 28-1, and starts removing the refrigerant. While the recovery process is taking place, Bob changes the air filters and oils the evaporator fan motor and the condenser fan motor. He examines the contactors and finds them to be in good shape, and he changes the indoor air filters as well.

Once the refrigerant is removed, he patches the leak using silver solder and repositions the piping so that it won't rub again. Then he pulls a vacuum on the unit and is ready to charge it with refrigerant.

BTU Buddy asks, "How are you going to charge the unit? There are no directions on it."

**FIGURE 28-1.** This recovery unit uses no oil and is very handy because it is light and you don't have to keep oil in the crankcase of its compressor.

"Let's use the superheat method like we used on the other system," Bob replies.

BTU Buddy says, "You can use a similar method, but remember, the TXV controls superheat, so it can't be used. The similar method would be to use the same instruments, the gauges and the temperature tester, but to charge for the correct subcooling in the liquid line. A good start would be to just add liquid to the liquid line until it stops flowing. This method won't overcharge the system but will add enough refrigerant that the system will function. Then we can add the rest as vapor into the suction line. When you start letting the liquid into the system, it will move toward the condenser and the evaporator. When the liquid stops flowing, there should be a balance of liquid in the system that distributes it in the condenser and evaporator. Then it will be ready to start."

Bob turns the R-22 cylinder upside down and adds liquid refrigerant. When it stops flowing, he turns the cylinder upright and closes both valves on his gauge manifold to start the unit.

BTU Buddy says, "Install a temperature tester lead on the liquid line and don't forget to insulate it so it won't register the surrounding temperatures. Also, remember that it's only 80°F outside, so the condenser air will need to be blocked to simulate a 95°F day."

Bob checks the head pressure and it's 210 psig, corresponding to an 80°F day (80°F + 25°F approach temperature = 105°F condensing temperature). The temperature tester is set up and the liquid line temperature is 100°F.

BTU Buddy says, "That's a subcooling of 5°F, and we haven't blocked the condenser yet to get the head pressure up to that of a 95°F day."

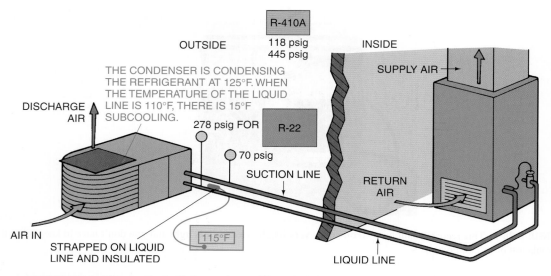

**FIGURE 28-2.** This standard efficiency air conditioner is being charged using the subcooling method.

Bob blocks the condenser, and the head pressure begins to rise. He adds vapor refrigerant in small amounts in the suction side of the system, until the liquid line reading is 110°F and the head pressure is 278 psig, corresponding to a condensing temperature of 125°F. The subcooling is now 15°F (125°F condensing temperature – 110°F liquid line temperature = 15°F subcooling), Figure 28-2.

Bob then asks, "What would happen if there were no subcooling and the liquid line temperature were 125°F?"

BTU Buddy says, "The system would operate and cool, just not up to capacity. More important than that, it wouldn't be as efficient. The customer may never know the difference in the efficiency, but if you left the system like that, you'd know that you didn't give the customer their money's worth. You're well-paid to provide professional service, and I know that's what you want to do. You can charge a system to 10°F and up to 20°F of subcooling, and it will be good."

As Bob puts away his tools, he says, "You really did show me how to do the right thing. Do you suppose all technicians try to do the job correctly?"

BTU Buddy says, "Many technicians don't appreciate the difference in getting the correct charge into the system, and some don't care—they're just out for the paycheck. I wouldn't be helping you if you didn't want to do the job correctly. It always pays to give the customer the best value for their money. I'll bet that your mother said something like, 'What you hand out is what you'll get back,' didn't she?"

Bob laughs, "Yes, she always told me, 'What goes around comes back around.' I suppose it's the same thing."

# Helping on a 100°F Day

The weather is hot—unusually hot. The design temperature for the area is 95°F and the actual temperature is 102°F. Bob gets a call to go to a car dealership late Friday afternoon, the complaint being that the showroom temperature is 85°F. They're having a big sale Saturday morning, and the showroom needs to be much cooler by then.

Bob arrives and is outside at the condensing unit, looking troubled, when BTU Buddy appears and asks, "What's the trouble, Bob?"

Bob says, "The owner of this business is really mad because it's Friday afternoon and he's having a big sale in the morning, and the showroom is 85°F. From what I see, the unit is doing all it can. I don't know what to do next."

"Well, an angry customer is no fun, that's for sure," BTU Buddy replies. "Here are some facts that we know of without even checking the unit:

• The outside temperature is 102°F and the unit was probably designed to hold the showroom down to about 75°F with an outside temperature of up to 95°F.
• The system uses R-22.
• The unit is probably doing all it can, but we should do a performance check on it to be sure.
• We can get the unit to perform a little better than its rated capacity for a short period of time.
• We can do some temporary things to reduce the load on the showroom tomorrow.

Let's run the performance check first."

Bob goes to the truck for his gauges, ammeter, and temperature tester and puts them in place. BTU Buddy has Bob fasten the temperature tester lead to the liquid line to check the subcooling.

Bob says, "Here are the readings:

- The suction pressure is 78 psig corresponding to 46°F boiling temperature in the evaporator. That seems about right because the temperature in the space is 85°F. At 75°F indoor temperature it would be about 40°F. It's hot in the conditioned space.
- The head pressure is a little over 300 psig. It's 102°F outside so we should be condensing at about 132°F. The head pressure is about right.
- The liquid line temperature is 130°F with only about 2°F of subcooling.
- The compressor is running at full-load amperage, which is 100 amps.

The unit is doing all it can."

BTU Buddy adds, "We don't know what the superheat is at the evaporator. The subcooling should be somewhere between 10°F and 20°F. We can get a little more from the unit by adding refrigerant."

"I'd be afraid to add refrigerant with the head pressure running that high," Bob says.

BTU Buddy replies, "The head pressure won't rise; we'll just get more capacity from the unit. The unit capacity will rise about 1% for each degree of subcooling until the subcooling circuit is full, then the head pressure will rise."

Bob charges the unit to 18°F of subcooling, and the head pressure remains the same.

BTU Buddy says, "Let's find out what the true superheat is at the evaporator. This will give us more accurate information than checking the superheat like we do with a residential unit. We'll fasten the temperature tester lead to the suction line at the expansion valve bulb location," Figure 29-1. "We could get a false reading if we placed the bulb after the external equalizer line," Figure 29-2.

Bob fastens the bulb to the line and insulates it, and the line temperature shows 56°F. He says, "The suction boiling temperature is 45°F and the suction line temperature is 56°F, so the superheat is 11°F. That's as close as we can get."

BTU Buddy agrees and says, "You have tuned the system so that it's doing all it can, using its air-cooled capacity. If you could lower the head pressure a little, we could get even more from the unit. Here's what we'll do:

- We'll set the thermostat to 70°F and turn off the nighttime timer that would shut the unit off at 9:00 p.m., at closing time. We'll let the unit have the space cool when we get here in the morning so it won't have to play catch up. The mass of cars and floor will already be cool. There are going to be extra people and more door openings tomorrow, so let's give the unit every advantage to work above its design capacity.

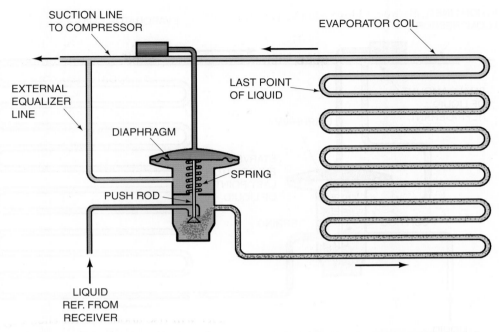

**FIGURE 29-1.** This illustration shows correct expansion valve bulb and temperature test lead location.

- We'll pull a water hose up here tonight, and on the way home we can stop at a store to get one of those soaker hoses that sprays a fine mist. We'll get a couple of flat pieces of wood and fasten the soaker hose to the wood, so that in the morning we can turn on the mist under the air-cooled condenser and lower the head pressure a little for even more capacity. This is a temporary setup for two reasons: it uses of a lot of water, and the water will leave mineral deposits on the coil fins. We don't want too much of that.
- We'll get a yard sprinkler and water the roof of the showroom to lower the solar heat load on the room.

These measures should get you through a bad day."

Bob and BTU Buddy arrive at the dealership with their sprinkler system and mist system at 7:00 a.m. and get set up.

The showroom temperature is 70°F at 8.00 a.m. when the doors open and people start to arrive. Bob turns on just the roof sprinkler.

At about noontime the showroom temperature rises to 73°F, and there's a crowd of people in the showroom. The head pressure begins to rise above 275 psig when the outside temperature goes above 95°F so Bob turns on the mister under the condenser. The head pressure drops back to 260 psig.

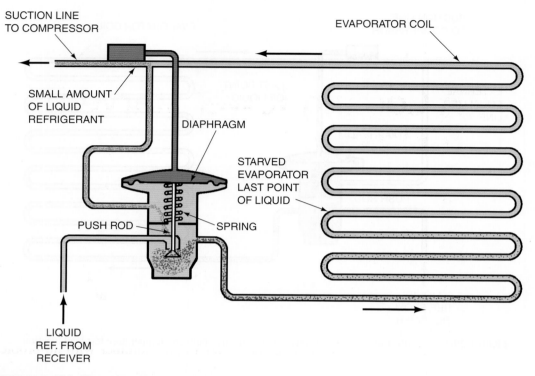

**FIGURE 29-2.** This illustration shows what can happen if the bulb and temperature test lead are placed at the wrong location.

At 4:00 p.m. the temperature in the showroom has risen to 75°F, but the crowd begins to leave. All is going well.

At 6:00 p.m. the last customers leave and the day is over. The showroom temperature never went above 75°F. The owner comes out to where Bob is and says, "You did a great job here today. The outdoor temperature went to over 100°F, and you were able to hold the showroom temperature to 75°F. I understand that you went to a lot of trouble to get this equipment to out-work itself, and I want to thank you for a professional job well-done."

Bob replies, "I've had a lot of help from the schooling and instructors that I had. My schooling experience was worth every bit of what it cost. Some of the tricks I learned from my instructor really pulled us through the day. Hopefully there won't be any more 100°F sales days. This weather is really out of normal range of what the equipment is designed for."

While riding away, BTU Buddy says, "The owner gave you a real compliment. It cost him to keep you here on overtime, but he probably made a lot of money today as a result of your help."

# A Cooling Tower Starved for Water

Bob receives a call from the dispatcher to go to an office building that has a 100-ton system that's water-cooled. The compressor is in an equipment room in the basement, and the cooling tower is on the roof. The building manager says the building temperature is rising. The outdoor temperature is 95°F.

Bob arrives and talks to the manager to see what he can learn. The manager tells Bob that system has been running all summer with no problem, but that lately the compressor has been shutting off because of high head pressure. He has been resetting the manual reset control several times each day, but today the building temperature continued to rise, so he called for service. This is the first time Bob's company has been called out to this job.

Bob and the manager go to the basement equipment room. The compressor is off again and the button on the high pressure control is out, signaling that the high pressure control had tripped, Figure 30-1. The manager reaches to reset the control and Bob says, "Let me look around for a minute before we reset it." The manager leaves the job to Bob.

Bob looks around the equipment room. The chilled water pump is running. The leaving chilled water is 75°F. No wonder the building is hot, he thought. The leaving chilled water should be about 45°F.

Bob moves to the condenser water pump and sees that it's running. The water coming from the tower is 85°F, which is about right, so Bob resets the high-pressure control and the compressor starts running. The compressor starts to load up toward full load and Bob thinks things are looking good, so he decides to just observe for a few minutes.

After about 15 minutes, the compressor tone begins to change; it sounds like it's straining. The head pressure would normally be running at about 210 psig for an R-22 system, water-cooled, on this type of day. The head

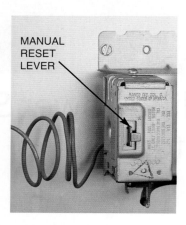

**FIGURE 30-1**. Manual reset high pressure control. The manual reset feature calls attention to the problem.

pressure is up to 250 psig and rising. The condensing temperature should be 105°F, corresponding to 211 psig, and is determined by the entering water temperature to the condenser, which is 85°F. Bob is scratching his head and wondering what to do when BTU Buddy appears and asks, "What's up, Bob?"

Bob explains what he knows about what's going on and says, "All looks normal, except the head pressure is rising."

BTU Buddy asks, "What is the leaving water temperature from the condenser?"

Bob looks and says, "It's 105°F. It should be 95°F. The condenser is taking a lot of heat out of the refrigerant. It seems like there's too much heat for the cooling water."

BTU Buddy asks, "What should the temperature rise be across the condenser? What is the temperature rise across *this* condenser? What would happen if there were plenty of heat, but not enough water?"

Bob says, "The temperature rise should be about 10°F," Figure 30-2. "The temperature rise across this condenser is 20°F. If there is not enough water flow, there would be an increase in temperature rise. I think we don't have enough water flow."

BTU Buddy says, "Let's go to the roof and look around."

Once they arrive at the roof, Bob says, "Boy, look at the steam rising off of the tower. That water is hot."

He looks into the tower basin and says, "The water level is low. The tower must be evaporating more water than is being made up and the water flow from the float valve seems low. I'll use that water hose to add water," Figure 30-3.

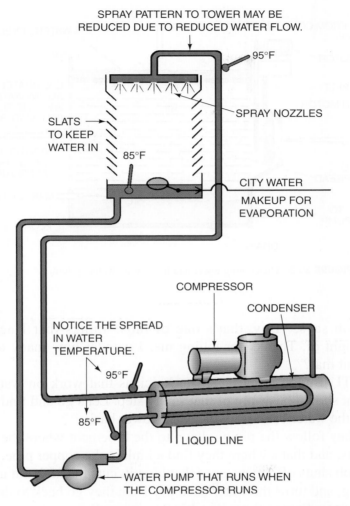

**SPRAY PATTERN TO TOWER MAY BE REDUCED DUE TO REDUCED WATER FLOW.**

95°F

SPRAY NOZZLES

SLATS TO KEEP WATER IN

85°F

CITY WATER MAKEUP FOR EVAPORATION

COMPRESSOR

CONDENSER

NOTICE THE SPREAD IN WATER TEMPERATURE.

95°F

85°F

LIQUID LINE

WATER PUMP THAT RUNS WHEN THE COMPRESSOR RUNS

**FIGURE 30-2.** This diagram shows the relationship of the cooling tower to the compressor and condenser when things are operating normally.

Bob brings a water hose over and starts adding water. As he starts to lay the hose over into the tower basin to let it flow, BTU Buddy stops him saying, "Tie the hose up high. Don't lay it in the basin."

"Why? What's wrong with that?" Bob asks.

BTU Buddy says, "If you were to walk off and leave the hose in the basin and the building lost water pressure, the water from the basin would be sucked into the building water system. The water in the basin looks clean, but you can bet it's not. Laying the hose in the basin has caused accidents where people became sick from drinking the polluted water. The water tower is actually a filter for whatever's in the air."

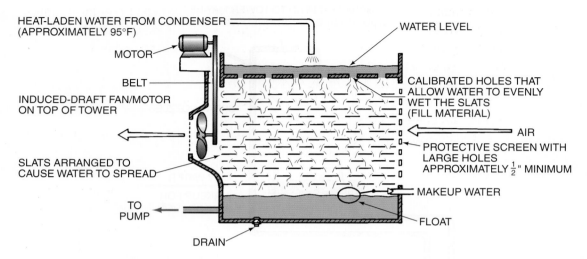

**FIGURE 30-3.** The cooling tower and its makeup water system.

Bob says, "Boy, that's one of those things that I never would have thought of. Thanks for telling me. I wonder how many technicians know about that?"

BTU Buddy replies, "All technicians that work on water-cooled equipment should be aware of that safety step. Let's go and find out what's holding the water back from the water tower basin."

They follow the piping down to the basement where the water fill circuit starts, and that's where they find a kink in the copper pipe.

Bob shuts off the water to that circuit, cuts the kink out and puts in a coupling, and turns the water back on. Then they go back to the roof to observe the water flow to the tower. It's flowing well.

Bob says, "Good thing that the water hose and the tower-feed water were not on the same circuit. The tower is nearly full now. Let's go and see what's happening."

Bob disconnects the water hose and they head back to the basement.

"Wow, look at that compressor run," Bob says. "The head pressure is 210 psig and the leaving chilled water is 48°F and on the way down. This job is fixed, thanks to you and another good experience."

Bob finds the building manager and explains what happened. The manager says, "Thanks for making a quick repair. The complaints have stopped coming in, and I've assured everyone that the system is working now."

Bob then says to the manager, "This system really needs regular maintenance. Would you like a quote for changing your filters, oiling the motors, checking belts, and keeping the cooling tower up to standards? You have a good system here, but it should be maintained."

The manager agrees, "You're right. I've been changing the filters, but the system would be better served by an expert. That will allow me to take care of the things that *I* am an expert in."

Bob says, "I'll turn the request in to the service department and they'll give you a quote. You'll hear from them in a few days."

As they're driving away, BTU Buddy says, "Great job on the service end, but also a great job in helping your company grow. Service contracts are good for the customer and for the company. When service calls are down in mild weather, the company can keep technicians busy with service contracts. Revenue is coming in for the company and the technicians have steady work."

# Evacuation Leak

Bob just changed a compressor on a 5-ton air-conditioner that cools the clubhouse at an apartment complex. He's just about ready to charge the unit when a problem arises. He has added a liquid line drier and sight glass just beyond the drier at the outdoor unit, and has leak-checked the connections he made and was satisfied there were no leaks. He has pulled a vacuum down to 500 microns on his electronic vacuum gauge, and is letting it sit for a few minutes when he notices a rise in the vacuum gauge to 1,000 microns. He's looking for an answer when BTU Buddy appears and asks, "What's the problem, Bob? You looked puzzled."

Bob describes what he has done, then says, "This system is 15 years old and refrigerant has never been added to it, according to its history. I don't get it; there shouldn't be a leak."

BTU Buddy asks, "Why did you change the compressor?"

Bob explains, "The compressor was just turning, not pumping, so I changed it. I'll need to start it up to explore what the problem could be."

Meanwhile, the vacuum gauge has climbed to 1,500 microns. There is a leak somewhere. Bob asks, "Could there be moisture or refrigerant in the system that's still boiling and making the pressure rise?"

BTU Buddy says, "It's not likely. Let's just let it sit for a few more minutes and see. If it's moisture or refrigerant still in the lines, it will stop soon. It's not likely that you could have lowered the pressure to 500 microns with water or refrigerant still in the system."

They wait for 30 minutes and Bob examines the system while waiting. He changes the filters and oils the motors, and while he's at the evaporator he finds a potential problem. But it doesn't seem to have anything to do with the possible leak.

Bob says, "The expansion valve bulb is loose on the suction line at the evaporator. It looks like someone just loosened it for no reason."

BTU Buddy explains, "Some older technicians who only go by the touch system of checking superheat may have loosened it to get more refrigerant flow, most likely while touch testing the suction line. The problem with that is that they wouldn't be able to tell how much more flow they created."

Bob says, "I'll bet that's what ruined the compressor. The oil in the crankcase was probably washed out and then the bearings seized and the crankshaft broke. You could hear the old compressor turn, but nothing happened. It had very low amperage when I checked it."

BTU Buddy says, "That's a good assessment of what may have happened. How does the vacuum look?"

"It's up to 2,500 microns," Bob says.

BTU Buddy says, "The vacuum rise tells you that you have a leak, but not where it is. Put a trace of R-22 in the system and push the pressure up to 150 psig with nitrogen, and let's look around. The first thing that you need to check is your connections. Remember, they were added to the system while you were doing your work. It's not uncommon for a gauge manifold to leak."

Bob turns off the valve to the vacuum gauge and adds the refrigerant and nitrogen, Figure 31-1. He then uses his leak detector to leak-check his own gauges and the connections he made for pressure testing, and he finds them

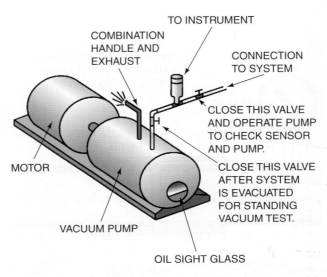

**FIGURE 31-1.** This valve arrangement can isolate the vacuum pump from the sensor and the sensor from the system.

all to be good. Then he starts to check the connections at the compressor and the drier and sight glass connections.

"Wow, there's a small leak around this sight glass connection that I don't believe was here before," Bob says.

BTU Buddy says, "It may not have been there when you leak-checked it. Notice that connection is made with 45% silver brazing rod because the pipe connection is brass, and you had to use flux to make the connection. Take a close look at where the leak is, and use soap to pinpoint the leak. Tell me what you find."

Bob looks closely and says, "It looks like the leak is under a bubble of flux."

BTU Buddy says, "That's exactly where it is. I noticed that you didn't chip away the flux after the connection cooled. There's a pinhole under the flux that may have been closed when you leak-checked it, and while working around the connection, you loosened the flux and it rose up off of the leak," Figure 31-2.

Bob lets the nitrogen and trace of R-22 out of the system and repairs the leak. This time he uses a screwdriver tip to chip away the flux after it cools. Then he pressurizes the system again, and the leak is gone. He lets out the nitrogen and starts the vacuum pump.

As the vacuum pump is running, Bob says, "I could have saved some time on that one. I can't really charge the customer for that mistake that cost about an hour of labor."

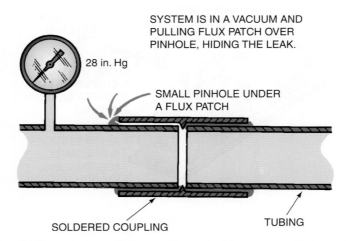

**FIGURE 31-2.** This illustration shows how a leak can be covered up by the flux. If the flux had not popped up with pressure on the system, it would have been sucked into the hole and would appear later during normal operation.

BTU Buddy says, "Well, some technicians would charge the customer anyway, but that's not treating the customer like they'd like and expect to be treated."

Bob shuts off the vacuum pump at 500 microns, and allows the system to sit for 15 minutes. There is no rise in the micron reading.

Bob says, "I'm going to charge the unit and consider it repaired after start-up."

BTU Buddy says, "Don't you think that you should check the superheat after start-up? You shouldn't just presume the valve is working correctly."

Bob charges the unit and checks the superheat at the condensing unit, and it's 15°F. The line set is about 35 feet long, so this looks about right. He says, "Well, another lesson that turned out okay. I didn't charge the customer for the extra time that I wasted in haste."

BTU Buddy says, "That's the best thing to do: be fair and honest."

# Working with a Gas Furnace that Is Down Drafting

The dispatcher calls Bob first thing in the morning before he leaves home and tells him that a customer with a gas furnace has noticed a strange smell in the house. Bob tells the dispatcher to call the customer back and have him turn the furnace off until he can get there, within the hour. It's very cold outside, about 10°F, but the customer has two fireplaces that can keep the house warm enough until Bob can get there.

Bob arrives and the customer explains that when the furnace is operating, he smells something peculiar, and he's concerned. It smells like his gas grill when it first starts up.

Bob turns the furnace on and goes to the furnace room in the basement of the house, where he finds that the furnace is operating correctly. It's an atmospheric draft system with a draft hood, as shown in Figure 32-1. He removes the burner compartment door and observes the flame. It appears to be burning properly; all blue and no visible yellow tips. Bob is thinking that there's nothing wrong with the furnace; there must be something else wrong. Maybe the homeowner has some other odor problem. BTU Buddy then enters the scene and asks, "Is the furnace drafting like it should?"

Bob says, "I don't know. All looks well."

BTU Buddy tells Bob, "Why don't you check the draft with a match, just to be sure."

Bob says, "We did this in school by lighting a match and holding it up close to the draft diverter to see which way the air was moving."

Bob holds the lighted match up close to the draft diverter and the flame moves away from it. Flue gases are not going up the flue. They're spilling into the basement instead, Figure 32-2.

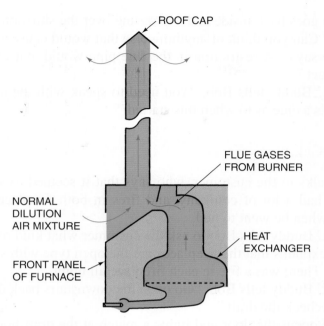

**FIGURE 32-1.** This is a typical natural draft furnace with a draft diverter operating correctly.

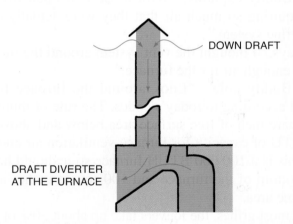

**FIGURE 32-2.** This draft diverter is not venting. This could be called back drafting or down drafting. The products of combustion are spilling out in the furnace area.

Bob steps out back to see what kind of chimney cap the flue has. It all looks normal. He suspected that a limb might have fallen on it and collapsed it, but it looks good.

Bob goes back inside and is puzzling over the situation when BTU Buddy asks, "Can you think of anything else that would cause a flue to not draft?"

Bob says, "A restriction in the flue pipe would, but everything seems to be intact."

BTU Buddy tells Bob, "You need to speak with the customer and see if there is a clue as to when this started."

## A CLUE ABOUT THE FLUE

Bob talks to the customer, who says that it seemed to start late last night. They had a lot of company, and fires in both fireplaces. He smelled the odor when he went to bed.

BTU Buddy tells Bob to ask the customer what kind of fireplaces he has.

He explains that the fireplaces are the open type with screens only and no glass. There was a fire in each fireplace all night.

BTU Buddy tells Bob, "Go open the downstairs back door about 4 inches and recheck the draft."

Bob opens the door and lights a match at the draft hood again, and says, "This draft is working now."

BTU Buddy explains, "With those two open fireplaces running, they were requiring so much air that they were actually drawing air down the furnace flue system."

Bob says, "I thought the vent system around the furnace was supposed to provide enough air for the furnace."

BTU Buddy notes, "Look around the furnace for the vents that are required according to today's codes. The rule of thumb is that there should be 1 square inch of free surface area below and above the furnace for each 1,000 BTU of capacity for furnace ventilation air and proper venting practices. This is a 100,000 BTUH furnace so it should have 100 square inches at the bottom of the furnace and 100 square inches at the top. Remember, this is free area.

"With most grilles, the louvers take up about 30% of the area. A grille with 100 square inches of free area would have an overall area of about 140 square inches. That is, a $10 \times 14 = 140$ total square inches; $140 \times 0.70 = 98$ square inches of free area. This means that a $10 \times 14$ grille should be located at the bottom of the furnace and at the top for proper ventilation."

Bob looks around and says, "There's not enough air vent surface area for this furnace. I can tell without even measuring. How did they get this past the building code?"

BTU Buddy responds, "In the early 1970s, homes were not as tight and were able to get ventilation air from the cracks around windows and doors. This house was built during that time, as you can see by the tag that's pinned up on the wall for the date of insulation."

Bob looks and sees that the home was built in 1970.

BTU Buddy continues, "The oil embargo in the early 1970s caused much stricter building codes in our part of the country. Builders began to tighten up all construction. This meant that mechanical ventilation codes had to be stricter. This is probably the original furnace, or the inspector would have required that proper ventilation be installed when the furnace was changed."

"What can we do now? Can we make the customer install the proper ventilation?" Bob asks.

BTU Buddy says, "No, you can't. But you can explain the circumstances, and the owner will probably want to remain safe and get the job done. For now, he can let those fires die down and leave the downstairs door open a crack so the furnace can get the ventilation it needs until the fireplace fires go out.

"Then the customer would be well advised to get glass doors for the fireplaces to prevent them from drawing so much heated air out of the house. An open fireplace draws air out, which pulls air in even when it's not burning. At the same time, you can give them a price on installing furnace ventilation. It won't be too complicated, because the furnace room is on an outside wall. Notice that the furnace room door has a louvered door. That door should be changed to a solid door to keep the ventilation air system within the furnace room. The louvered door was installed originally to pull air from the whole house for ventilation air. This system should have been corrected years ago.

"You can see that someone has installed storm doors and windows all around the house. The windows have also been caulked to tighten the house against outside air infiltration. All of this contributed to the problem today. This house is really tight compared to when it was built."

Bob asks, "How much danger was the family in with those fumes?"

BTU Buddy explains, "The fumes that they smelled were the aldehydes that are present in all gas burning products of combustion. These are present in flue gases from good combustion and poor combustion. Poor combustion occurs when there's not enough oxygen to properly burn the fuel; yellow flame tips are the indication of poor combustion. With good combustion, you can still smell the aldehydes, but the gas is not poisonous. It will suffocate you because of lack of oxygen, but it's not poison, like carbon monoxide. It's not likely that suffocation would occur, as the oxygen level in the whole house would have to be reduced to a very low level. Most houses would leak in enough air to prevent suffocation. Proper combustion gives off carbon dioxide. Both smell about the same and are dangerous, with carbon monoxide being poisonous. The people were in no immediate danger, but it was not a good situation.

"This house should also have a carbon monoxide detector, as all homes should. I would suggest that you draw up a proposal to install the ventilation system, and encourage the customer to have fireplace fronts installed along with two carbon monoxide detectors, one for upstairs and one for downstairs. It's a good idea for all technicians to carry a carbon monoxide detection instrument for situations just like this. We were able to look at this job and find a solution easily because of past experience."

Bob says, "The experience was all yours. I'm still learning. I'm glad that all of my service calls aren't this complicated. Most of them are fairly easy."

BTU Buddy says, "Most calls are routine. The more experience you have, the more routine calls you have. This is a profession where you never stop learning, which is part of what makes it so interesting. There are always challenges out there to keep you thinking. This service call was one that allows you to become a safety inspector for the customer. I think that all customers would welcome a chance to upgrade their system to make it safer. You aren't just selling them more service; you're providing them peace of mind.

"When houses were built back in the late sixties and early seventies, no one knew that the energy crunch would come and that houses would be modified to become tighter. Infiltration was not even considered back then. All construction was very loose. The construction of today creates houses that must have planned ventilation air systems."

# *33*

# Handling an Over-Fired Boiler

Bob is notified by the dispatcher that the owner of a small office building has called to say that his oil boiler needs a tune-up. It hasn't been serviced in about two years and frigid weather is expected.

Bob drives up to the building and finds the owner to talk to him, inquiring about the previous servicing of the boiler. The owner explains that various people have worked on it, and that the last few times, his brother-in-law did the work. He wasn't sure how much his brother-in-law knew about boilers; he's a sort of jack-of-all-trades. He has since moved to another state, so it's time to really take a close look at the boiler.

Bob goes to the basement where the boiler is located. He looks around for about 15 minutes and goes back up to talk to the owner, and asks him to go to the boiler room with him. When they get there, Bob explains, "Before I can go to work on this boiler, there must be some preliminary work done in this room. There are papers, mops, brooms, floor sweep compound, and various other combustible materials in the room that must be removed. It's a small room and a fire could start easily with all of this material lying around. I'm surprised the insurance inspector hasn't found this and required that it be cleaned up."

The owner responds, "You're right. A fire could really be a problem. This is awful. Can you come back after lunch? By that time I'll have this cleaned up."

Bob says, "Okay, please have this room in good shape so I can get started."

Bob writes up a report and files it in his paperwork, just in case.

Bob returns after lunch to find that the room has been cleaned up, swept, and mopped. He's ready to go to work now. The first thing he does is shut off the power to the boiler and lock out the panel. He shuts off the fuel line from the tank, which is above the burner. Then he takes a small pan and places it under the fuel filter and removes the filter. It is very dirty, but still working. He changes the filter and fills the filter cartridge with fuel oil and puts it back together.

He then goes to the boiler and opens the inspection door to the combustion chamber. He uses a flashlight to inspect the combustion chamber with a mirror that has a telescoping handle made for this type of inspection, Figure 33-1. The combustion chamber looks good. He looks at the front of the burner head at the same time. It looks good, although there is some varnish on the burner head, like it has been running hot.

Bob raises the burner ignition transformer and removes the nozzle and electrode assembly. Here is where some questions start to pop into his head. The nozzle has a heavy coat of varnish on it for no apparent reason. He looks at the nozzle size and angle, and replaces it with an exact replacement. He uses a wrench that is specifically designed for removing nozzles, Figure 33-2. This wrench holds the nozzle holder while he loosens the nozzle, which is the best method to prevent bending the electrodes while removing the nozzle.

Then he cleans the electrodes and insulators to remove any carbon residue. This will prevent the high voltage from arcing to ground on the carbon. He uses his gauge to set the electrodes.

The nozzle assembly is now ready to set back in place. Bob takes one more step, he pours the nozzle piping full of fuel oil and puts his thumb over the end of the pipe as he slides the assembly back together. Filling the piping helps to prevent an air bubble from getting into the piping. An air bubble can be very hard to remove. You would think that the high pressure oil would push it out, but it doesn't always, and can cause after-fire drip, which is noisy and creates soot.

While Bob has the transformer up, he removes the cad cell and cleans it with a soft rag. This cad cell must be able to see the combustion when the

**FIGURE 33-1.** A telescoping mirror used by a heating technician.

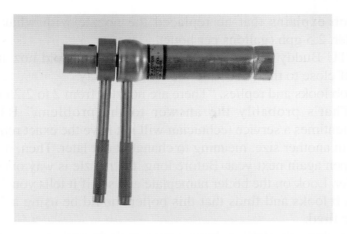

**FIGURE 33-2.** An oil burner nozzle wrench.

burner light's off. He is careful not to put his fingers on the contact prongs at the back of the cad cell; this could cause a poor connection in the electronic circuit.

After he fastens the transformer back in place, he is ready to start the boiler and take an efficiency test. He installs a draft gauge over the fire through the hole in the inspection door. There is no plug in the hole; Bob makes a note to install one. He turns on the fuel valve and the power. The boiler fires correctly and looks great. He looks through the inspection door at the flame; it looks big, but seems normal for now.

Bob installs a thermometer in the stack in preparation for an efficiency test. The temperature is going up. As a matter of fact, it should stop rising and it doesn't. It is running entirely too high. The stack temperature went right on past 800°F and kept going up. Bob notices that the metal pipe from the boiler to the masonry chimney begins to glow dark red. About that time, the boiler shuts down; the thermostat was satisfied.

## TOO HOT FOR COMFORT

Bob is looking puzzled when BTU Buddy appears and says, "It seems that too much heat is going up the flue. What was the draft gauge reading?"

Bob says, "Things were happening so fast that I forgot to look."

Because it's cold outside, the boiler starts back up in a few minutes. Bob looks at the draft gauge and it's –0.03 inches wc (inches of water column). This is fairly normal. The draft in the flue is –0.07 wc. The temperature is rising again.

BTU Buddy says, "What size oil nozzle did you use?"

Bob explains that he replaced the nozzle with what came out of the boiler, 2.5 gph (gallons per hour).

BTU Buddy says, "Look around at all of the old nozzles that are on the shelf close to the boiler and see what they say."

Bob looks and replies, "There are nozzles from 2 to 2.25 to 2.5 gph here."

"That's probably the answer to the problem," BTU Buddy says. "Oftentimes a service technician will not have the exact replacement and will put in another size, meaning to change it out later. Then, if he forgets, it may happen again next year. Before long, the nozzle is way out of match with the boiler. Look on the boiler nameplate and see if it tells you."

Bob looks and finds that this boiler should be using a 2 gph nozzle. It is over-fired.

He shuts the boiler down again and changes to the 2 gph nozzle. He writes on the side of the boiler what the correct nozzle characteristics are. He then picks up all of the old nozzles and throws them away so the same mistake won't be made again.

BTU Buddy then suggests, "While the system is off, install a pressure gauge in the pump discharge. Someone may have changed the pressure setting."

Bob installs a high pressure gauge and starts the boiler again. The oil pressure is 110 psig. Bob says, "No wonder the flue was getting too hot. The nozzle was oversized and the pump was putting out too much pressure. Why didn't a safety control shut the boiler off?"

BTU Buddy explains, "The boiler thermostat took care of it. The boiler water never actually got too hot, just the flue. This was a very inefficient operation, with a lot of heat going up the flue. You didn't do an efficiency test before changing the nozzle, but I'll bet it was below normal. The brother-in-law was not a real technician."

Bob runs a flue gas analysis test on the boiler. The flue gas temperature is running 650°F, a much more acceptable temperature. He then does a smoke test that shows the flue gas is clean. The boiler efficiency is running at 71%. This has to be much better than when they arrived, not to mention the safety factor.

Bob asks the owner to come back down to the boiler room and he explains what went on. Then he says, "I would suggest that you get only qualified service technicians to do your work. This was a very dangerous situation. With all the combustible material that was in the boiler room when I arrived, this could have been a major liability for you."

The owner says, "You don't have to say any more. You will be our technician from now on."

While cleaning up the tools, BTU Buddy says to Bob, "You're turning into a first-rate technician, Bob. There are several things you did that showed real respect for the customer, which will aid you and your company over the long haul:

- You used a pan to catch the dripping oil when you took the system apart. Just a few drops of fuel oil can cause a callback. The customer might have thought there was a leak.
- You locked and tagged the electrical circuit while working on the boiler to prevent a mistake. If you had left for a part during this time, someone could have turned the boiler back on without knowing, which could have caused damage.
- You observed the signs of over-firing before you actually found it. Always look for clues.
- You cleaned up the tracks of the other technicians and left a trail as to what was correct by indicating the nozzle size on the side of the boiler. Many service people will forget to look on the boiler nameplate.
- When you removed your instruments, you put covers over the holes.
- You got the owner involved, which he appreciated.

"Bob, you are becoming a great technician; just keep learning and pass on everything you know to your fellow technicians. It will come back to you in the best of ways."

# Tackling Low Airflow with Electric Heat

Bob gets a call from the dispatcher about an office that has electric heat. It's very cold outside and one of the office zones is not heating as it should. It's 62°F in that zone.

Bob arrives to find that the zone temperature is now about 60°F, and the personnel are very uncomfortable. Small electric space heaters have been brought in to spot heat the zone that is so cold.

Bob goes to the equipment room and finds a blueprint of the building, and discovers that there are about 10 electric heaters that have a wiring diagram like that in Figure 34-1. Each zone has an air handler, so they are all the same.

Bob goes to the air handler for the zone that has the problem. It's above the ceiling, so he sets up a ladder. There's a light bulb next to the air handler, so there is plenty of light. He can hear the fan running, and the duct seems to be hot, so something is working. All looks well, so he is standing on the ladder looking bewildered when BTU Buddy appears and asks, "What's the problem?"

Bob says, "There are so many wires in this panel that I don't know where to start."

BTU Buddy says, "Let's do it like all problems; see what we know, then explore what we don't know, and find out the answers. First, identify the low voltage circuits entering. It is that terminal block that's isolated from the rest. Remember, low-voltage must be isolated from high-voltage. Identify the high-voltage terminals that connect directly to the heater circuits. Now, take your ammeter and check the amperage to each heater."

Bob takes an amperage reading at each heater. He discovers that some have current flow and some do not. "Maybe this is why there is not enough heat," he says.

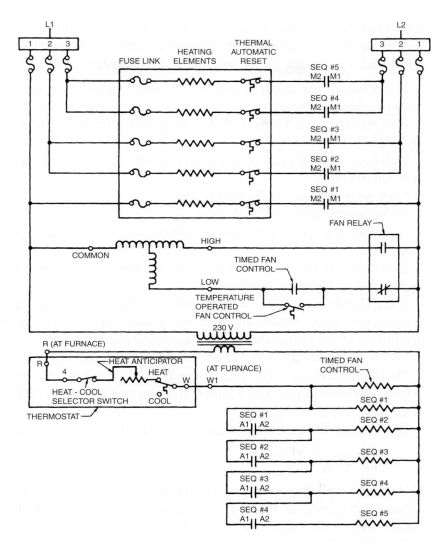

**FIGURE 34-1.** This wiring diagram shows five stages of electric heat.

BTU Buddy suggests, "Bob, take the amperage at both sides of each electric heat strip. Make a drawing of what the amperage is at each place."

There are five heat strips that are 5 kW each, so Bob takes 10 readings. The voltage is 230 volts. The amperage should be about 22 amps at each place (5,000 watts ÷ 230 volts = 21.7 amps). The results look like this from the top to the bottom, using Figure 34-1.

First heater is 22 amps on the left and 22 amps on the right.
Second heater is 18 amps on the left and 0 amps on the right.

Third heater is 0 amps on the left and 0 amps on the right.
Fourth heater is 22 amps on the left and 22 amps on the right.
Fifth heater is 0 amps on the left and 15 amps on the right.

Bob looks at the readings and says, "Now I'm more confused than ever. What is going on here?"

BTU Buddy says, "This will all add up in a minute. We'll get to the bottom of it."

Bob says, "I don't see how a heater can be drawing current on one side and not the other."

"Just wait and see," BTU Buddy says.

"What do we do next?" Bob asks.

"Let's start by pulling the heater section out of the furnace," BTU Buddy suggests.

Bob shuts off the power and locks out the power supply, and with the key in his pocket, he feels confident.

BTU Buddy says, "Before you put your hand in that electrical panel, check the voltage. Fasten one lead to a ground and touch your other meter lead to all of the terminals and look for any signs of voltage."

"With the power off, why do I need to do this?" asks Bob.

"It's the safe thing to do. There have been cases where circuits from other sources have been run into a panel, and you don't want to find this out the hard way," BTU Buddy says.

Bob checks all circuits to ground and finds the panel safe to work in.

## CHECKING OUT THE HEATER SECTION

Bob loosens the power wiring so that he can pull the heater section out. Then he disconnects the low voltage power supply. He pulls the entire heater section out and lays it on the floor for examination.

He says, "This heater was really hot at one time. Look at the elements. Some of them have burned in two and are touching the frame."

BTU Buddy says, "Get your diagram of the heaters with the amperage results you recorded, and list what you see beside each heater."

The following is what Bob noted:

First heater is 22 amps on the left and 22 amps on the right. **This heater looks good.**

Second heater is 18 amps on the left and 0 amps on the right. **The heater coil is burned in two and one side is touching the frame.**

Third heater is 0 amps on the left and 0 amps on the right. **This heater has a burned fuse link.**

Fourth heater is 22 amps on the left and 22 amps on the right. **This heater looks good.**

Fifth heater is 0 amps on the left and 15 amps on the right. **The heating element is burned in two and one side is touching the frame.**

BTU Buddy asks, "What conclusions can you make from this?"

Bob says, "I can understand why the number 3 heater is not drawing current. It needs a new fuse link," Figure 34-2. "But I don't understand heaters 2 and 5. Why are they drawing current on one side only?"

BTU Buddy explains, "There is still power to the heating elements on both sides, but the coil is burned in two. One side is suspended in air, while the other is touching the frame and has about 115 volts across that part of the coil," Figure 34-3.

"Well, now I need to go to the supply house for parts," says Bob.

"What are you going to get?" asks BTU Buddy.

"I need two heating elements for this specific heater, and a fuse link," Bob says.

BTU Buddy suggests, "Check the continuity across the automatic reset thermal controls on the units that were not working, just in case one of them has malfunctioned."

Bob does this and finds that one of them indeed has an open circuit. He finds the rest to be good.

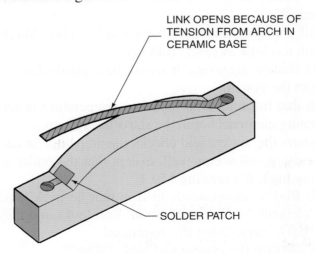

FIGURE 34-2. This is a drawing of a fusible link.

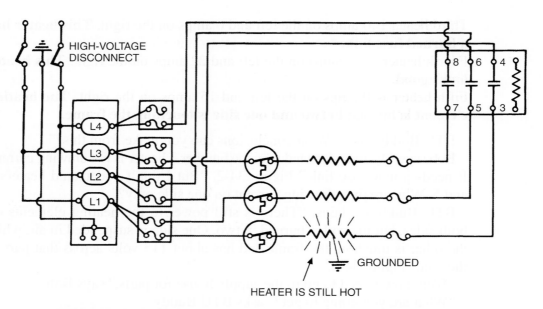

**FIGURE 34-3.** This shows a heating element that is grounded to the frame and still heating.

Bob picks up the parts he needs and arrives back at the job. He installs the new heating elements, automatic reset thermal control, and fusible link. He is ready to put the system back into operation when BTU Buddy asks, "What do you think happened to this system?"

Bob says, "I have no idea. It probably just burned up."

BTU Buddy says, "You can do better than that. Did you see any signs of problems?"

"Well, the system looked like it had been hot. Maybe it had been operating with too little airflow," Bob says.

BTU Buddy suggests, "It would be a good idea to check that out when you start the system."

With that in mind, Bob locates a temperature tester lead downstream of the heating elements before he starts the system.

He starts the system and checks the amperage at each heater, and they're all operating. All seems well, except that the outlet air temperature seems to be too high. It's running 150°F.

BTU Buddy recommends to Bob that he peek inside the duct and look at the elements while in operation. Bob looks and they're glowing red-hot. BTU Buddy says, "Shut the system off."

Bob shuts off the system and asks, "Why?"

BTU Buddy says, "These elements are not meant to glow red-hot. There's not enough airflow across them, for some reason, and they're going to burn out again. It's not worth taking the risk of letting them run."

"Well, what do we do?" asks Bob.

BTU Buddy says, "Let's look for a restriction in the airflow."

Bob takes the panel off of the air handler and looks at the fan and the air-conditioning coil. Now he sees what the problem is. The coil is matted with dirt on the inlet side.

Bob asks, "Now what do we do? It would be a big problem to clean this coil above the ceiling with detergent and water."

BTU Buddy says, "Look at the fan wheel. It's really packed with dirt. This system must be cleaned before it's put back in operation. Notice that the fan is after the coil in the air stream. If the fan wheel is this dirty, you can imagine what the inside of the coil must look like because the dirt on the fan wheel is what came through the coil. This probably happened during the summer, when the coil was wet. This job must be done correctly or there will be problems in the near future."

## CLEANING THE FAN WHEEL AND COIL

Bob gets set up by shutting off the electrical power and locking it down. He then removes the fan wheel. He carries it outside and sprays detergent on it to let it soak. He has to place it in a parking garage to keep it from freezing.

He goes back in and spreads plastic on the suspended ceiling under the fan coil unit. He cleans the entire matted surface from the coil inlet with a vacuum cleaner to prevent it from having to go down the drain when the coil is washed. He then uses a pump-type sprayer and sprays the coil with an approved coil cleaning liquid, paying close attention to getting the core of the coil wet. He sprays both the front and back of the coil. He then goes to the outside to clean the fan while the coil soaks. The fan wheel cleans up easily with the detergent.

Bob installs the fan wheel after giving the coil another soaking. He then uses a pressure washer to clean the coil. He says, "I can't believe all of the dirt coming off the coil. There must be a problem with the filtering system on this unit."

BTU Buddy says, "Now you're thinking, Bob. When you get it cleaned up, we'll look for the 'why' of what happened."

Bob had to really watch the condensate drain line that was draining off all of the water and detergent, to make sure it didn't overflow, even though the system had a secondary drain pan.

With the coil clean and the fan wheel cleaned and back in place, Bob starts the fan only, and lets it run for a few minutes to prevent water from blowing into the electric heat coils. Then he turns on the electric heat and measures the air temperature with all of the heat running. The air temperature is running at 115°F. All seems to be well.

BTU Buddy says, "Let's get to the root of this problem and see why that coil was so dirty."

Bob removes the filter and it's dirty, but not plugged completely. There's a lot of dust in the return air duct that seems out of the ordinary. He goes to the conditioned space where the return air is entering the system.

"I see what the problem is," says BTU Buddy. "What do you see, Bob?"

Bob says, "Boy, I don't know. There seems to be a lot of dust in this room but I can't imagine why."

BTU Buddy says, "This is the mail-opening room. A lot of dust is created by mail, but even more is created when the mail is opened. Notice that there's also a shredder. That also creates a lot of dust."

Bob asks, "Why doesn't the filter remove that dust from the air?"

BTU Buddy explains, "Typical media filters are not that efficient; some people say about 5%. They pick up large particles but not much fine dust. So most of the fine dust, such as in a mail room, will go right through. It takes a while, but eventually enough dust will lodge in the coil to restrict the airflow, especially during the cooling season when the coil is wet."

"Well," Bob says, "another new experience. Does it ever end?"

"Let's hope not," says BTU Buddy. "All of these new experiences are what keeps this profession interesting. My granddad from Alabama used to say that everyday is like knots on a pine log; each day, like the log, has a different pattern."

He continues, "Bob, you did a good job on this. I think you should explain the situation of the high dust level to the management here, and recommend that a higher density filter be installed, and that a routine maintenance contract be established to keep all of the systems in this building in good shape."

Bob returns after talking to the building manager with a big smile. The manager asked for a quote for the contract, and assured Bob that he would take it.

BTU Buddy says, "Now, see? You've brought your company some more business. You are becoming more valuable each day to them. Service contracts are great because you can work them on 'off days' when there are no service calls. It provides a set amount of revenue that the company can rely on each month and year. Great job, Bob."

Bob then asks, "How do all of those thermal overloads work together with the fuses to make a system reliable?"

BTU Buddy says, "Meet me for lunch tomorrow with a pad and a diagram of a system, and we'll discuss it in detail. You must understand the entire system, not just one or two controls. They must all work together."

# A Lunch Seminar with Bob and BTU Buddy

Bob and BTU Buddy have met at the local diner for lunch. After their meal, BTU Buddy says, "Now that we've had lunch, let's have a cup of coffee and talk about electric heat and the components that operate it."

"Where do we start?" Bob asks.

BTU Buddy says, "The first thing you must realize is that a lot of energy is dissipated with electric heat. The reason for this energy release at the heaters, instead of where the wires enter, is management of electrical resistance in the electric circuit."

Bob asks, "What do you mean by management of resistance?"

"Well," BTU Buddy says, "have you ever wondered why the electric heat element gets hot, yet the wires leading up to it don't?"

Bob says, "I don't think I ever thought of that, but you're right. The heat is concentrated at the heating element."

BTU Buddy explains, "The reason is that the wires leading to the heating element have very little resistance to the flow of electrical energy. The heating element itself is made of a substance called nichrome, which stands for nickel-chromium, an alloy. When electrical energy of a known potential is applied to the element, it will cause electron flow through a controlled resistance, and heat is given off. This heat is transferred to the air passing over the elements. As you discovered on your last service call, if there is not enough air flowing, the elements will get too hot. Too much heat remains in the coils. They can overheat and burn up. This is the same process that takes place in a lightbulb, except the lightbulb is contained in an enclosed atmosphere with no oxygen. The element can glow 'white hot' in this atmosphere and not burn out. This is not so in the oxygen atmosphere of air.

"Now, let's look at the controls that you asked about while we were servicing the electric heat unit the other day. There are some variables that must be controlled with electric heat. If there is a misapplication or a reduced airflow to the point of danger to the system, it must be shut down.

"There were two kinds of protection to the heating elements in the service call we were on. The diagram you have will reveal these and the reasons for them," Figure 35-1.

BTU Buddy continues, "This circuit can be seen as containing three different characteristics: the power supply (L1 and L2), the conductors

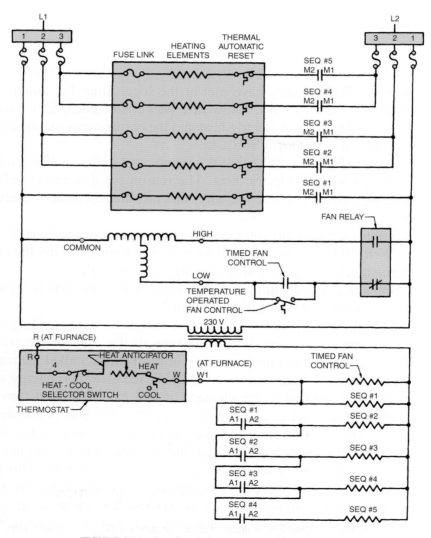

**FIGURE 35-1.** An electric heat system wiring diagram.

(the wires leading up to the heaters), and the load (the actual heaters). Looking at the right side of the heater box diagram, we see a thermal overload, or limit control, represented by a switch that opens with temperature. This is often called a snap disc," Figure 35-2. "This control may be set to

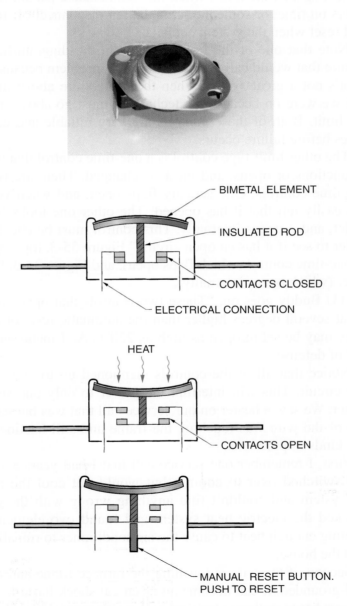

**FIGURE 35-2.** Automatic and manual reset limit controls that may be used for electric heating systems. The diagram illustrates manual reset limit control operation.

open the circuit at 180°F and close at approximately 150°F. This is the first line of defense against overheating. These temperatures sound high, but remember that they are looking straight at the heating element and will have some influence from radiant heat at that location. These controls will protect against the times when the maintenance person doesn't change the filters on time, or someone turns the fan disconnect off for a minute. They will reset when the system cools.

"Note that one of the limit controls in the diagram has a manual reset feature that would call your attention to a problem because it must be reset. That's not a great system when the heaters are above the ceiling like the call we were on because the technician must go above the ceiling to reset the limit. Both of these controls are very reliable and can function many times before failure occurs.

"The other limit-type control is a one-time control that will not reset once it functions or opens, and must be changed. There are two types of these one-time controls. One actually flaps open, and when you look at it, you can easily tell that it has opened. The other one looks like a small silver bullet, and it opens internally. This control must be checked with an ohm-meter to see if it has an open circuit." Figure 35-3, top, shows the first type of one-time control (which flaps open), and Figure 35-3, bottom, shows the other (which opens internally).

BTU Buddy goes on, "These two controls that open and don't close are set at several degrees higher than the automatic reset or the manual reset. They may be set to open as high as 220°F. As I mentioned, this is the last line of defense.

"Notice that all of the controls mentioned up to now actually open the line circuit. This will interrupt the power to only one side of the electric heater. We saw a heater on our service call that was burned in two, and one end of the wire was touching the frame and still heating. This can create two kinds of problems.

"First, I remember one service call that I had years ago where a system had switched over to cooling but would not cool the house. I checked the system and couldn't find anything wrong with the cooling. I finally checked the electric heat system and found two elements grounded and creating enough heat to cause the air conditioner to run all the time and not cool the house.

"Second, if the wire is touching the furnace frame and the furnace is not well-grounded, it can become an electrical shock hazard. Any time you are under a house with an electrical system it's a good idea to push one lead of

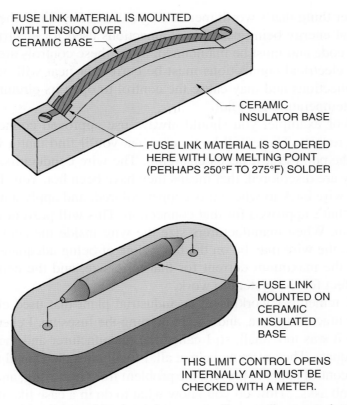

FUSE LINK MATERIAL IS MOUNTED WITH TENSION OVER CERAMIC BASE

CERAMIC INSULATOR BASE

FUSE LINK MATERIAL IS SOLDERED HERE WITH LOW MELTING POINT (PERHAPS 250°F TO 275°F) SOLDER

FUSE LINK MOUNTED ON CERAMIC INSULATED BASE

THIS LIMIT CONTROL OPENS INTERNALLY AND MUST BE CHECKED WITH A METER.

**FIGURE 35-3.** Two types of fuse links used for electric heat systems. One actually opens and can be seen. The other is enclosed and must be checked with an ohmmeter. These types of limit switches must be replaced once they function.

your electrical meter into the ground and touch the other to the appliance. If there is a voltage reading, shut the power supply off using your insulated screwdriver handle. What you've done with the meter is simulate what would happen to your body if it were to become an electrical path to ground. On wet ground, you don't have a chance if the frame is electrically hot.

"We've discussed the thermal safety control, often called the limit controls. But we haven't discussed the fuses in the system. The fuses are protection to the conductors—the wires servicing the heaters. They will also protect them from a direct short to ground, but not the type of ground you had on your service call. That ground had the resistance of the heating element still in the circuit."

Bob says, "There are a lot of safety controls in an electric heat system."

BTU Buddy agrees, "Remember what I said earlier. There is a lot of energy potential in the area of an electric heat system, and there is one other thing that's worth mentioning at this time. Since there is a lot of electrical energy being used, the connections used for electric heat must meet the code and must be maintained. Since these controls are sensitive to heat, the electrical connections must be correct or heat will be generated at the connections and may cause the control to open its circuit, even though the air temperature isn't hot.

"For example, you should always use approved connectors when making repairs. There will be times when you'll find that a wire has been hot to the point that it needs changing. The wire conductors are copper and if they are discolored, that means they have been hot. You should cut the copper wire back to where it is copper colored, and apply a new, fresh connector that's approved for that connection. This will prevent it from happening again. When manufacturers size the wire inside the actual heater terminal box, the wire may be on the borderline of being adequate. It may be carrying the maximum current for that wire size, and the connections must be perfect for the system to work.

"I remember working in an industrial plant that used electric heat in the manufacturing area, and it kept burning the fuses up. I checked the wire size and it was too small, so I called the manufacturer and they said that it was adequate. They were no help at all. I finally installed small fans in the electrical control panel to prevent the problem from occurring, and it worked."

Bob asks, "How do you know what to do in a case like that?"

BTU Buddy says, "It was a matter of excess heat, and the manufacturer didn't want us to rewire the electrical panel. So, we did the next best thing; we added ventilation. This is actually the job of the electrical department, but they had exhausted their means to make a repair and turned it over to us, as it was a heating system."

Bob asks, "Will I ever be able to have that kind of judgment?"

BTU Buddy answers, "Yes, because you keep asking questions and reaching for solutions. You are on the right track—just don't get derailed."

# A Problem of House Windows Sweating

The first cold weather of fall has prompted a service call to a residence because all the windows in the house are sweating on the inside. The customer explained that all seemed well with the heating system until the windows began to sweat, and he wanted to know if the heating system could have anything to do with the problem.

Bob arrives and talks to the homeowner, who tells Bob that he just wants to make sure the heating system is working properly. For some reason the windows began to sweat when the weather became cold. It all seems to relate to turning on the heating system.

Bob goes to the partial basement where he finds a gas furnace. He starts the furnace and everything looks good. It's an older natural draft furnace, and the flame is good and blue and seems to be burning well. He holds his hand in front of the draft diverter to see if the flue gases are going up the flue, and they appear to be, but he decides he will do the candle test that he learned in school to be sure, Figure 36-1. He brings back a candle from his truck and holds it in front of the draft diverter. The candle flame is pulled into the diverter, so the furnace is venting correctly. Bob is about to go to the homeowner and declare that the furnace is performing fine when BTU Buddy makes an appearance and says, "You know the furnace is venting correctly now, but maybe you should try some other tests and ask some questions."

Bob asks, "What kind of tests and questions?"

BTU Buddy says, "Didn't we have a service call last winter where the furnace was down drafting because there were open fireplaces that pulled air out of the house?" (See "Service Call 32: Working With a Gas Furnace that Is Down Drafting.") "What would happen if the owner were operating

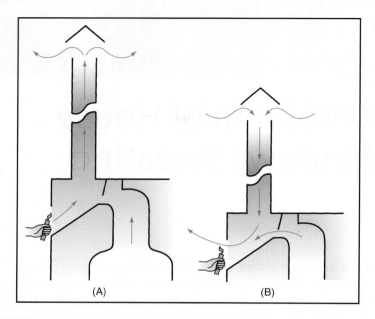

**FIGURE 36-1.** Furnace draft test. (A) illustrates furnace up draft. (B) illustrates furnace down draft.

several exhaust fans at the same time? What if the vent system has a loose connection between the furnace and the chimney and all of the flue gases are not going up the chimney? Any of these would cause flue gases to move into the basement area and filter upward into the house. Remember that a large part of the flue gas is moisture. You may also want to look at the clothes dryer vent to be sure it's venting outside like it should. It's obvious that the air is loaded with moisture, because the windows are sweating on the inside. You must look for any source of extra moisture."

Bob says, "Boy, there sure is a lot to think about. This could be a complicated service call."

BTU Buddy says, "It could also be simple. You just have to do your investigation. Talk to the homeowner about some of these items that we discussed."

Bob goes to the homeowner and asks, "Have you been using your fireplace much this year?"

The homeowner says, "No, we haven't had a fire in the fireplace at all this year."

"Have you had any reason to run the exhaust fans a great deal?" Bob asks.

The homeowner says, "No, we cook with gas and run the kitchen exhaust when we cook, but we've been eating out for the past two weeks because of a busy schedule."

Bob then asks if he can see the clothes dryer. They go to the basement and Bob sees that the dryer is a gas unit. He thinks he has found the cause of the moisture so he asks the homeowner to turn the dryer on for a running test. He goes outside to where the vent terminates, and it's venting fine. The vent only goes through the wall.

Bob thanks the homeowner and goes back to the basement, scratching his head with more questions than answers.

He goes to the furnace for a look at the entire flue system. The flue is piped with single-wall vent pipe. This would not meet today's codes, as it's an unconditioned space. The vent leaves the back of the furnace and is piped through the crawl space, but he can't see where it enters the chimney because it goes around a corner. Bob puts his coveralls on and crawls along the path of the vent. When the vent goes around a corner to the other side of the chimney, he finds the problem. The vent pipe had been jarred loose and is venting out into the crawl space. He looks closely and there are no screws in the vent pipe. When it was assembled, the installer did not fasten it correctly.

Bob comes out from the crawl space and BTU Buddy asks, "What did you find?"

Bob explains what he found and tells BTU Buddy that he's going to fasten the single-wall pipe with screws, as it should have been in the first place.

BTU Buddy tells Bob, "That will be a repair, but it won't bring the venting system up to the latest code standards. Why don't you explain the non-code situation to the homeowner and give him the option for a repair or replacement vent pipe."

Bob fastens the pipe together with screws. He also looks at all the connections and makes sure they are all fastened correctly. This puts the system back in good working order. He then goes to the homeowner and explains the situation.

The homeowner says, "It has worked well all of these years. Why should I pay to change it?"

Bob explains, "The furnace will continue to vent correctly as it is, but the single-wall vent will sweat and eventually deteriorate. The code was changed to prevent that, making systems safer."

The homeowner says, "Go ahead and get your installation crew to change it to meet the code, and thanks for doing a great job of explaining the system to me. When will my windows clear up from the sweating?"

Bob responds, "It will take a few days, but it will gradually go away. We found the source of the moisture and stopped it."

As Bob and BTU Buddy are riding away, Bob asks, "Why didn't the homeowner smell the gas fumes?"

BTU Buddy explains, "The furnace was burning good and clean, and the products of combustion have a very faint smell. I smelled it when we went into the house. Even a properly burning gas flame gives off a light smell because some of the products of combustion contain aldehydes, which are in the same family as formaldehyde. It's not easy to smell, but once you've smelled it and know what it is, you would probably notice it the next time."

BTU Buddy tells Bob, "You did a good job with that customer. When you make the customer part of the call, they have more confidence in you and your company."

# Finding Excess Heat Exchanger Scale

On a very cold day, the dispatcher has called Bob and sent him to a home where there is no heat. Bob arrives at the job and meets the homeowner. She explains to Bob that the furnace operated all night and stopped running in the early morning. She leads Bob to the basement where the furnace is located.

Bob checks the thermostat wires and sees that the thermostat is calling for heat, but the furnace is cold. He removes the cover to the burner section of the furnace and discovers that the burner section has a lot of soot where one of the burners is inserted into the heat exchanger. They are ribbon-type burners, Figure 37-1. The flame rollout protection control has tripped and needs to be reset, Figure 37-2. When the burner starts up, the flame is yellow and rolling out the front of the burner hole in the heat exchanger. Bob tells the owner that the burners will have to be removed and the furnace cleaned before start-up.

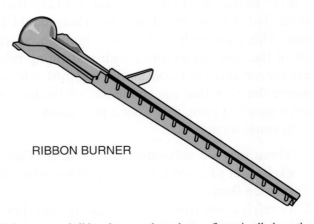

RIBBON BURNER

**FIGURE 37-1.** This is a stamped ribbon burner where the gas flame is all along the burner top.

MANUAL RESETTABLE
FLAME ROLLOUT SAFETY

**FIGURE 37-2.** This manual reset flame rollout switch can save the furnace wiring by shutting off the flame if for any reason the flame rolls out the front of the burner hole.

Bob turns off the gas and power to the furnace and begins to remove the burners. When he removes the burner that has the soot buildup, he notices some large pieces of scale lying on the burner surface, Figure 37-3. Bob is starting to think that something is really wrong at this point. He's never seen this before. BTU Buddy shows up about this time and asks, "What's got you puzzled, Bob?"

Bob says, "That's a large patch of rust lying on the burner. I wonder if the heat exchanger is damaged?"

BTU Buddy explains, "It's more likely that it's just surface rust and scale, but it bears checking out. There could be several reasons for that rust. This furnace has electronic ignition, so the pilot light doesn't burn all of the time. Sometimes with a standing pilot, that portion of the heat exchanger stays hot constantly, and rust will accumulate above the pilot where the pilot flue gas condenses on the heat exchanger. But since this is not a standing pilot, that's not a possibility.

"It could be that:

1. There may be solvents stored nearby. They will often cause the heat exchanger to corrode in large patches when the fumes are burned by the furnace flame.
2. If refrigerant from a long-lasting leak has been introduced into the burner section, it will corrode the heat exchanger.

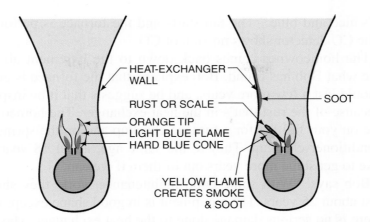

**FIGURE 37-3.** A rust patch that has fallen on the burner head will cause the flame to impinge on the heat exchanger and burn yellow, giving off soot.

3. If the basement where the furnace is located is damp, rust may eventually find a place to form under the coating on the heat exchanger.

Use those as a checklist for excess heat exchanger scale, and see what you find."

After removing the burners, Bob takes them out to the truck where he can blow them out with an air tank. The burners are in good shape when he's finished cleaning them. He takes them back in and uses his flashlight to look into the heat exchanger before starting to clean it. He says, "There seems to be a lot of soot in this one section where the scale was found." He removes the draft diverter so he can see down into the heat exchanger. There is some soot in the draft hood.

He gets his shop vacuum from out of the truck and cleans the draft hood. He then uses a long brush to clean down into the heat exchanger from the top, and also to clean up from the bottom. He scrubs the heat exchanger until no more soot is falling down, then he vacuums out the heat exchanger. He again takes his flashlight and examines the heat exchanger very carefully. Then he says, "There seems to be more rust in this one section of heat exchanger, but they're all rusty to some extent. The heat exchanger seems to be solid, with no holes. What do you suggest?"

BTU Buddy suggests, "Put it back together and start it up, and see if the flame is stable. It would be a good idea to place your CO (carbon monoxide) detector above the heat exchanger to see if there is any CO present, However, I don't think you'll find any."

Bob assembles the furnace and puts his detector probe above the heat exchanger and starts the furnace. He remarks, "The flame looks good.

It's nice and blue." The fan starts and the furnace is performing correctly. The CO detector shows no sign of CO.

The homeowner comes back down to the basement about this time to see what Bob has found. Bob explains that the furnace is getting older, but should last a few more years, and he suggests that it be inspected every fall because of the rust that's in the heat exchanger. The homeowner says, "Put me on your fall list for a furnace checkup and on the spring list for an air-conditioner checkup. They're the same age, about 18 years old, but we'd like to get some more years out of them if we can."

Bob says, "With some good maintenance work, they should be able to last about 25 years. The equipment is in good shape except for the rust, and there is no serious damage done to the heat exchanger. May I ask, does the air-conditioning system require that refrigerant be added very often?"

The homeowner says, "Yes, we had refrigerant added every year until this past spring when one of your technicians found a leak on this line, here at the furnace."

Bob responds, "That explains a lot. When refrigerant escapes near a flame, some of it will be drawn into the flame. This refrigerant turns into a mild acid when burned and that's what has caused the excess rust on the heat exchanger."

Bob puts his tools away, and as they're driving away, he tells BTU Buddy, "It's really nice to find the answers to these puzzles. A little knowledge and experience go a long way. Thanks for your help!"

BTU Buddy says, "Just keep learning and keep asking questions. Remember, you have two ears and one mouth. You can really learn more with your ears than your mouth. You did a great job with that customer. That question about refrigerant leaks was excellent. You also have great textbooks that can help you with a lot of these problems, as well. The burner in that furnace was a rather old-type burner. The new end-shot and up-shot burners don't protrude as far into the heat exchanger and are not subject to that kind of problem. Time and technology are marching on to produce new and better parts."

# Handling a Frozen Outdoor Coil on a Heat Pump

The day has been very cold and damp when a customer calls the dispatcher and says that the coil on her heat pump is frozen solid. The fan is making a noise, which called her attention to the frozen coil. Bob tells the dispatcher, "Call her back and tell her to switch the thermostat setting over to emergency heat." Bob has other calls to make before he'll be able to get to the frozen heat pump.

When Bob arrives, the customer meets him at the door and tells him that it's getting cold in the house. She asks why the emergency heat didn't heat the house up to temperature.

Bob explains that the emergency heat, plus the heat pump, are calculated to heat the house on a 20°F day, but the emergency heat switch on the thermostat only shut down the heat pump to keep it from doing damage to itself. It also turned on the auxiliary heat, which helps some, but won't keep the house very warm. Bob tells the owner, "If you want to bring the heat up a little, you should go to the kitchen and put a pan of water on each burner of your stove. Boiling the water will add humidity to the house, to improve the comfort inside. Make sure the pans have water in them to keep from overheating on the stove." She goes off to do this, as there is an older person in the house, as well.

Bob walks around the house to the heat pump and looks at it. There is a solid sheet of ice on the coils. He knows what to do first. He removes the top of the unit, and from the fan blades he clears away the ice that had been causing the noise the customer described. He then removes the control panel cover to get ready for a forced defrost. He goes to the thermostat and starts the unit, Then he goes to the unit and starts the defrost cycle at the electronic circuit board. He hears the four-way valve change over to defrost, so he just watches as the defrost cycle progresses. After about

10 minutes, the defrost termination timer stops defrost. Bob takes a close look at the coil and sees that not much of it defrosted. So, becoming more curious, he starts another defrost cycle. It doesn't seem as though there is enough heat to defrost. He's looking puzzled when BTU Buddy appears.

"You look confused," says BTU Buddy.

"Well, I guess I am. It will take all day for this heat pump to defrost the ice off of the coil, at this rate. Something is not right," says Bob.

BTU Buddy advises, "Turn the heat pump off with the breaker, and then you go to your truck for gauges and your leak detector. By the time you get back, you'll be ready to install them."

Bob has learned to follow BTU Buddy's lead, so he turns off the heat pump and returns with the gauges. When he starts to install them, BTU Buddy says, "Be sure to use the gauge ports that will give you the true pressure readings during both cooling and the heating cycle. Those are the ports that are on the outside of the cabinet, up on the side."

"Why did you suggest that we turn the pump off at the disconnect?" Bob asks.

BTU Buddy explains, "You could have gone into the house and turned the unit back to emergency heat, but then when you installed the gauges, you would have had to go in again and turn it back on. You can control all of that from here with no trips to the house. The reason for shutting it off was to let the pressures equalize. I suspect there's not enough refrigerant in the system to produce enough heat for a proper defrost. If you install the gauges while in a vacuum, you may create another kind of problem—you could pull air and moisture into the system. It's always better to install gauges when there is a positive pressure. Also, you should test the gauge ports before you remove the caps because the leak may be at the cap. Once you remove it, you have changed the conditions and you won't know if it was leaking when you arrived."

## FINDING A LEAK

Bob leak-checks around the gauge ports and finds that one of them is leaking. It's the high-pressure port in the heating cycle, which would explain why it had leaked down so quickly.

BTU Buddy says to Bob, "This system could have been low on refrigerant for quite a while, in which case the thermostat would start the auxiliary heat to help it when needed, and the homeowner may not have known the difference. Before we leave, you should explain the auxiliary heat light on the thermostat to her."

Bob then says, "We've found the leak, or at least one leak. I think we should start the heat pump and charge some refrigerant into it."

BTU Buddy says, "That sounds good to me. It is about 35°F outside, so what do you think the approximate suction pressure should be for this unit?"

Bob says, "I don't know. I know that the suction pressure should correspond to a temperature much colder than the 35°F outdoor air temperature, but I don't know how much colder."

BTU Buddy explains, "The actual suction pressure depends on several things: the outdoor air temperature, how clean the outdoor coil is, and the amount of moisture in the outdoor air passing over it. The suction pressure for this unit will be lower than normal until we get the ice melted off of the coil. I would suggest that we use a water hose to melt the ice, rather than multiple defrost cycles. I think it will be faster. Get some plastic to cover the electrical box and the compressor terminal box, *and turn off the power,* and I'll show you what I mean."

Bob brings some plastic sheets from his truck and covers the electrical boxes. He then turns on the hose and lets the water flow over the coil. He comments on how quickly the ice begins to melt.

BTU Buddy remarks, "The water from the city water mains is probably about 45°F and will melt ice quickly. The ice that's built up is not actually solid ice. It has a lot of air in it and will melt quickly."

When the ice has melted off, Bob starts the unit with his refrigerant cylinder connected to the system. The unit uses R-22. The suction pressure begins to drop, approaching a vacuum.

"Shut the unit off, and let's repair that Schrader valve that's leaking before we go on," BTU Buddy says.

Bob shuts the unit off and asks BTU Buddy what he suggests.

BTU Buddy says, "The Schrader valve service connection has two chances to prevent a leak—the seat in the actual valve, and the valve cap or cover. This valve has to be leaking from both the seat and the cover. Replace the seat under pressure using that special tool that you have," Figure 38-1, "and then replace the cap, and all will be fine."

Then BTU Buddy explains the answer to a previous question about suction pressure. "This unit seems to be a standard-efficiency heat pump. I believe the suction pressure should correspond to a coil boiling temperature about 25°F lower than the air temperature. That would be 10°F (35°F air temp − 25°F = 10°F). When we look at the pressure-temperature chart for R-22 at 10°F, we find that the suction pressure would be about 33 psig," Figure 38-2. "This is only an approximation; we'll need to go to a charging chart to fine-tune the charge. Use the chart that's inside the panel on the unit."

**FIGURE 38-1.** This valve can be attached to a Schrader port and the valve core can be removed and valved off. A new core can then be inserted, and the system can be put back in operation. All of this can be done under pressure without loss of charge.

Bob asks, "Why does the chart call for the outdoor wet bulb temperature as one of the pieces to the puzzle?"

BTU Buddy answers, "The outdoor wet bulb temperature is an indication of the total heat that is being absorbed into the unit. Remember, the outdoor coil is the evaporator during the heating cycle. The moisture on this coil is usually in the form of ice in the winter cycle and the load on the coil can vary by a considerable amount due to the differences in moisture in the air."

Bob uses the chart and wet bulb thermometer along with his gauges to bring the charge up to the correct operating conditions.

The homeowner comes out of the house about this time, and asks how things are going.

Bob shows her where the leak was and tells her that he has repaired the leak and everything is normal now.

She says, "I noticed that the system was really putting out a lot of heat and was hoping all was well. Thanks for the prompt and professional service. It was really getting cool in the house."

Bob replies, "You have a small blue light on your room thermostat that will light up from time to time. Have you ever noticed that light being on?"

"Yes, but I don't know what it means," she says.

Bob explains, "When that light is on, it tells you that the auxiliary heat is helping your heat pump to heat the house. Auxiliary heat is expensive to operate and should only be on in weather of about 30°F and lower. If you ever see that light to be on a great deal when the weather is 30°F or higher,

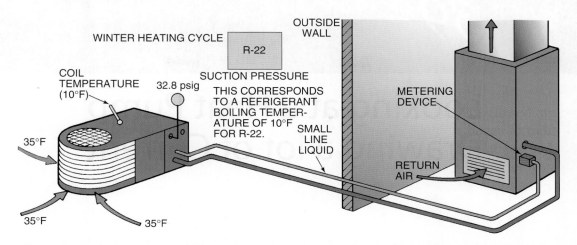

**FIGURE 38-2.** This standard-efficiency heat pump outdoor unit has 35°F air passing over the coil, and the refrigerant is boiling at 10°F, creating a suction pressure of 32.8 psig. The coil is boiling at 25°F lower than the outdoor air.

it's a sign that the heat pump is not carrying the load, and there may be a problem."

"Well, that makes sense," she responds. "I have noticed it to be on a great deal the last few weeks. I'll know what to do next time, thanks."

Bob gathers his tools and, as they're driving off, BTU Buddy says to him, "That went smoothly. You did a good job with that call."

"Why didn't you suggest that we recover the remaining refrigerant and charge the system from a deep vacuum, like many manufacturers recommend?" Bob asks.

BTU Buddy explains, "This was a much shorter process. It didn't involve the use of a recovery setup and evacuation along with scales to recharge the unit. It was less expensive, and frees you up to go on to another call."

Bob agrees, "Saving time is really important because of the workload that we have. And time saved for the customer is money saved for the customer."

BTU Buddy says, "Good point, and believe me, the customer notices an efficient service technician. That will lead to more business for you and the company."

# Looking at a Heat Pump Drawing a lot of Current

As Bob is getting ready to start his day, the dispatcher calls and reports that a regular customer wants their home heat pump checked out. The power bill has gone up dramatically in the last 2 months.

It's 28°F outside. Thank goodness it isn't snowing, Bob thinks. He arrives at the house and talks to the wife. She shows him her power bill and it is at least 40% higher than the month before. She says that the house has been comfortable, but that the light on the thermostat seems to stay on a lot more. In fact, it's on now. Bob walks down the hall for a look at the thermostat and sure enough, it's on. He goes outside to the outdoor unit and looks around. The large copper line going to the house seems normal, and there is a little ice on the outdoor coil. He goes to the basement and looks at the indoor unit. All seems normal there. He removes the electric heat access panel and uses his ammeter to check the electric heat. Two out of three heating elements are drawing current.

Bob decides to fasten gauges to the unit to see what the high- and low-side pressures are operating at. The unit has R-410A as the refrigerant. The evaporator in the outdoor unit should be boiling at about −7°F, which is 35°F less than the 28°F outdoor temperature. The suction pressure should be about 41 psig. He installs the gauges and discovers that the suction pressure is 62 psig, which corresponds to about 10°F. The suction pressure seems too high. The head pressure should be about 340 psig and it's 275 psig. He takes an amperage reading and discovers that it's lower than normal.

Bob is scratching his head over these figures when BTU Buddy appears and asks, "What do you think the problem could be, Bob?"

Bob says, "I think the compressor is not pumping correctly. The suction pressure is high, the head pressure is low, and the amperage is low. That sounds like an inefficient compressor to me."

BTU Buddy says, "If this were a cooling unit, I'd agree with you. But we have another factor in this unit; we have a four-way valve. You may need to test it. You don't want to change the compressor and find that you have the same symptoms when you start the unit with a new compressor. Let's check the four-way valve for leak-through next."

Bob turns off the outdoor unit and installs thermometers for checking the four-way valve. Figure 39-1 shows the arrangement for checking the performance of a four-way valve using temperature comparison. Two thermometer leads are needed, one on the cold gas line from the outdoor coil, and the other on the cold gas line going to the compressor. (The object is to measure both cold gas lines to see if hot gas is leaking into the suction line.) Bob then turns the unit back on and lets it run for about 20 minutes. Meanwhile, he checks the filter at the air handler and finds it to be clean and freshly changed. The wife comes down and tells him that she changes the filter every 60 days and that they're never dirty.

Bob goes out to the outdoor unit to check the temperatures. Since the suction pressure is 62 psig for the boiling refrigerant of 10°F, the suction line from the evaporator should contain about 10°F of superheat. So the line should be about 20°F, and it's 23°F—very close. However, the suction line leaving the four-way valve going to the compressor should be no higher than 3°F more than the evaporator suction line. It's reading 50°F. Bob says, "This four-way valve is really leaking through, causing the system to be inefficient. It's not the compressor."

BTU Buddy responds, "Now you're getting the picture. The valve is definitely defective. This would cause the auxiliary heat to operate for longer periods and cause the power bill to rise. We can try reversing it several times to see if we can free it up. It may have a piece of trash under the seat or it may have been damaged when it was installed. It isn't likely that it would be worn out because it doesn't get much activity."

Bob uses the outdoor wiring to cause the four-way valve to change over several times, and then he checks the temperature again. It's still leaking through, so he decides to tell the homeowner that the valve must be changed. The unit is only 8 years old and in very good shape, so changing the unit is not the best economical solution for the customer. He gives her an estimated price and she says to change it out.

BTU Buddy tells Bob, "Drop some of that ice from the coil onto the discharge line from the compressor, and see what happens."

Bob drops a little ice onto the line and it sizzles and boils. "What was that all about?" he asks.

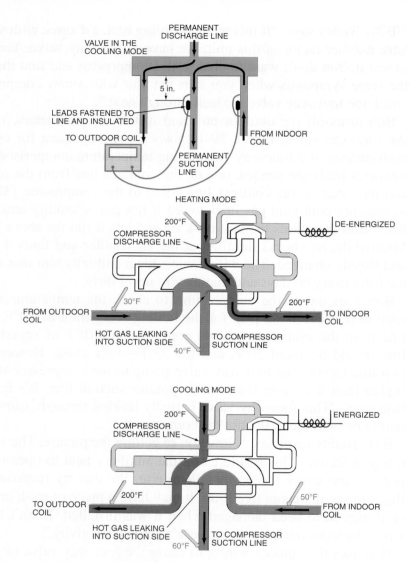

**FIGURE 39-1.** The upper illustration shows the arrangement for checking the performance of a four-way valve using temperature comparison. It indicates the correct placement of the temperature probes. Notice they are insulated. The two lower illustrations show how to check the valve in heating mode and in cooling mode.

BTU Buddy explains, "That compressor is running too hot because of the leak-through. It's probably shutting off because of internal overload from time to time. Under normal conditions, the discharge line should never be hot enough to boil water like that. Have the owner switch the unit to emergency heat until you can change the valve to prevent compressor damage.

When a compressor runs that hot, it will cause the oil to decompose over time. Let's not take a chance with this compressor; it's still running."

Bob arranges to change the valve out the next week because a break is predicted in the weather.

As they are riding away, BTU Buddy says, "Changing a four-way valve is a big and intricate job. I'll meet you here next week and walk you through the procedure."

**Note:** *In the next chapter of the BTU Buddy series, Bob and BTU Buddy install the new four-way valve.*

# Changing a Four-Way Valve on a Heat Pump

To finish the job they were working on last week (see "Service Call 39: Looking at a Heat Pump Drawing a lot of Power"), Bob meets BTU Buddy at the job site to change the heat pump's four-way valve on a cool but dry day. It's a good day to work outside. Bob knows that it's best to only open a refrigeration system on a dry day, and that it's much better when it isn't too cold, so that it's comfortable to work outside. BTU Buddy asks Bob what he thinks the best procedure would be to change the valve.

Bob says, "We'll have to recover the refrigerant, then turn off and lock out the power supply, and then remove the top of the unit so we can get to the valve. We should then break the vacuum after recovery with dry nitrogen, and use the multi-tip torch to remove the piping from the valve. There are four pipes, which we'll need to do one at a time. Then we can install the old piping to the new valve, in reverse. Is that what you would do?"

BTU Buddy says, "There are several ways of changing the valve. This valve is in a close space, and doing all of that torch work won't be easy. I think the preferred method would be to cut each pipe as close to the valve as possible, using a close-radius tubing cutter. I've always found it easy to remove the piping after melting it away from the valve. But when it comes time to push those pipes back into the new valve, those silver-based solder beads on the piping won't allow you to just push the pipes into the new valve," Figure 40-1.

"There are two ways that technicians deal with this. One is to file the hard solder off of the ends, though it's very hard to accomplish that all the way around the pipe. The pipe must be filed to the original size for it to fit the new valve. Another way is to simply heat the pipe and the valve connection and slide them together. Oftentimes this is very hard to do without overheating the valve, which has a fiber seat inside that can be ruined with excess heat. Melting the piping solder back into the valve is difficult and requires some real expertise. The third, and I think preferred method, is to cut the pipe back like I mentioned. Then put couplings in the pipe with short stubs of piping. There are more connections to solder, but you'll be soldering fresh connections that are much easier to work with," Figure 40-2. "You can do all of the fitting of the pieces outside the unit. You can even solder-in the stubs outside the unit if you want, but it's just as easy to do it all at once within the unit."

Bob cuts the old valve out of the unit. Three of the lines on the valve are 7/8-inch diameter and the discharge connection is 5/8-inch diameter. Next he places short stubs of pipe in each valve connection, with a coupling on each.

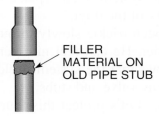

FILLER
MATERIAL ON
OLD PIPE STUB

**FIGURE 40-1.** Notice the silver solder beads on the piping. These are very hard and not easily removed.

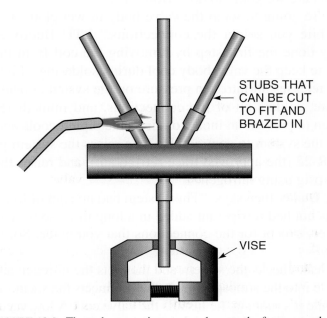

STUBS THAT
CAN BE CUT
TO FIT AND
BRAZED IN

VISE

**FIGURE 40-2.** The stub connections are made up to the four-way valve.

(He has already carefully cleaned the connections on the valve with approved sand tape, holding each one down so no sand residue enters the valve body.) When he has the connections on the valve made up, he cleans the four piping connections on the unit. He reams the connections inside, where he cut the pipe. He then places the valve and stubs into the piping in the system. Everything fits great.

BTU Buddy suggests, "You've had the piping open for a fairly long time. Don't you think you should install a two-way drier in the liquid line?"

"Good idea," Bob says. "I think you're right."

Then BTU Buddy says, "It always pays to keep a system extremely clean. Especially since heat pumps operate below freezing, it pays to keep them dry. You have the valve ready to braze in place. Next, use dry nitrogen to flush out the system, then replace the liquid line drier and you'll preserve the capacity of the drier."

Bob lets the nitrogen trickle slowly through the system, and installs the flare connection drier. He can feel nitrogen inflating both connections to the drier, so he knows that the system is clean and dry. Now he's ready to solder-in the four-way valve and stubs.

BTU Buddy says, "Let's protect this four-way valve from heat using an old-fashioned method. Get a small pan of water and some cloth strips from your truck."

"What are we going to do?" Bob asks.

"We're going to wrap the valve body in wet cloth strips and keep them wet while you solder the connections," BTU Buddy explains. "You've already done the first step by removing the coil from the valve. The next step is to keep the valve body cool during soldering," Figure 40-3.

Bob releases all nitrogen pressure on the system, opens it to atmospheric pressure, solders all of the connections, and immediately reconnects the nitrogen to prevent an intake of air as the piping cools and causes the nitrogen in the system to contract. He then raises the system pressure to 20 psig using R-22 (the approved trace refrigerant), and raises the system pressure to 150 psig using nitrogen.

BTU Buddy then says, "This system had no sign of low refrigerant charge and has not had refrigerant added in a long time, so we can assume that it is leak free, except for the connections that you made. So, leak-checking will be easy."

Bob leak-checks the system and then lets the nitrogen and trace refrigerant dissipate into the atmosphere. He then connects the vacuum pump and turns it on. While it's running, he installs the valve coil. A low vacuum (300 microns) is obtained in about 30 minutes. Bob says, "That really was fast."

**FIGURE 40-3.** This valve is wrapped in wet cloths to prevent the valve body from getting too hot during soldering.

BTU Buddy explains, "You took good precautions working with nitrogen, and there was no moisture in the system. You also used large vacuum pump connections, which cost a little more, but really speed up the job and pay for themselves in time saved."

Bob uses scales to measure the correct charge back into the unit. He starts the unit, and then uses the "O" wire that controls the four-way valve, to switch the system back and forth from heating to cooling several times, to make sure the valve is functioning properly. He removes his gauges and replaces all the panels, and then he goes to the room thermostat and sets the system back to normal heat.

As they are riding away from the job, BTU Buddy asks, "What did you learn from this job?"

Bob says, "That there can be more than one way to do a job. I liked using the stubs on the four-way valve to protect the valve body. The water and cloths were a great idea. I know there are other heat sink paste-type materials, but this proves that the job could be done correctly without them if they're not available. I also learned that using nitrogen while making high-temperature connections makes for a much cleaner job, and protects the system."

BTU Buddy says, "These are old-fashioned solutions that have worked for years. Many technicians are careless with moisture and air in a system. It always pays to do your best work. A technician should treat every system like a low-temperature refrigeration system, in order to ensure a long and healthy life for that system."

# Checking a Heat Pump with Dirty Indoor Coil

The first call of the day is from the owner of a strip mall. There is a post office in the mall that is serviced by a 10-ton heat pump. The owner told the dispatcher that the post office had called and complained about the electric utility bill for the last 2 months. It seems that it's much higher than usual.

It's a cold day for early spring—40°F—so Bob will be able to operate the heat pump under heating conditions without overheating the conditioned space. The indoor coil is in an equipment room with plenty of space around it, and the outdoor unit is on the roof.

Bob arrives and goes to the branch manager of the post office to find out what he knows. The manager shows Bob the electric bills for the winter and Bob agrees that they have increased significantly. The manager says, "I track the bills here very closely. I've been the manager for 10 years and I have never noticed a rise like this in the cost of power for a month. The lights are almost a constant, and stay on the same amount of time each month, so the heating system must be the problem."

Bob agrees and tells the manager that he'll get started, and see what he can find. He goes to the equipment room where the indoor coil is located. The first thing he does is check the filters; they're clean. He notices a log of filter changes attached to the unit. The filters have been changed frequently and the date noted.

The unit is running, and continues to run as Bob looks around. He uses his ammeter to check if any of the electric heat is operating. One stage is running, which doesn't seem right to him. He's sure the heat pump alone should be able to carry the load. He begins to think that there must be a control problem causing the electric heat to come on. Then he thinks the heat pump may not be operating to full capacity. He reaches over and touches the hot gas line and it seems too hot. About this time BTU Buddy appears and asks, "What seems to be the problem? You look confused."

Bob says, "This is a funny set of symptoms. The electric heat is operating and it's 40°F outside. The heat pump hot gas line is really hot. I don't seem to get the picture here."

BTU Buddy says, "How does the liquid line feel?"

Bob touches it and says, "Wow, it's really hot, too. That doesn't seem right either. I checked the filters and they're clean. The fan seems to be running normally. This just doesn't add up."

Then BTU Buddy asks, "Why is the liquid line so hot? What would cause that?"

"Usually reduced airflow, because the condenser is in heating mode," Bob answers.

BTU Buddy says, "Correct. Is the airflow really up to normal?"

Bob says, "I don't know. I'll do some checking around."

## CHECKING AIRFLOW

Bob goes to his truck and brings out a velometer, as seen in Figure 41-1. He checks the airflow at several registers and remarks, "The air velocity seems a little low. According to the chart, the velocity could be as high as 1,500 fpm (feet per minute) for this application, and it's running about 300 fpm," Figure 41-2. "I think we should look in the coil and fan section for a possible problem. This is a belt drive motor. Maybe the belt is slipping."

Bob shuts off the unit and removes the fan compartment door. He checks the fan belt and it's adjusted correctly. Then he starts to examine the coil. Using a flashlight, he looks at the coil from the outlet side and says, "This coil looks really dirty down in the core. I think it needs cleaning."

BTU Buddy agrees and says, "The key to solving this problem was the hot liquid line. You didn't even need to put gauges on the system to know that the air wasn't removing enough heat from the coil. Any time the liquid line is much warmer than your hand temperature, there is a problem with a heat pump. The liquid line should not be any warmer than 100°F. Hand temperature is normally about 91°F, so if the line feels really warm or hot to the touch, there's a heat transfer problem with the indoor coil. This is not true for an air-cooled condenser in the cooling mode, even on a heat pump. The air-cooled condenser may have 100°F air or higher passing over it. The return air for a heat pump is normally not warmer than 75°F, a much cooler condensing medium. Now what are you going to do?"

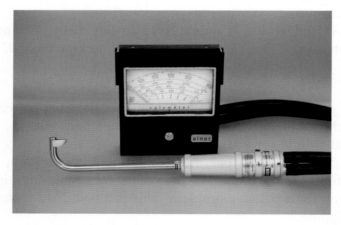

**FIGURE 41-1.** These two velometers are examples of those that are used for measuring air velocity.

Bob says, "I'm going to clean the coil and fan wheel."

"How are you going to clean them?" BTU Buddy asks.

Bob explains, "I'm going to turn off the power and lock it out, spread plastic over all electrical boxes and connections, and then I'll take the fan wheel and motor out. Then when I clean the coil, I can wash it counter to the airflow to push all dirt out the way it came in."

BTU Buddy says, "That's the correct procedure. You're getting good."

Bob removes the fan and motor and takes them outside for cleaning. The fan looks dirty, but not excessively. Then he sprays the fan wheel and coil down with approved detergent. He really gives the coil a good soaking and lets it get wet all the way through. After the detergent has had time to soak through the dirt, Bob uses a pressure washer to clean the coil and drain pan.

| Structure | Supply Outlet | Return Openings | Main Supply | Branch Supply | Main Return | Branch Return |
|---|---|---|---|---|---|---|
| Residential | 500–750 | 500 | 1,000 | 600 | 800 | 600 |
| Apartments, Hotel Bedrooms, Hospital Bedrooms | 500–750 | 500 | 1,200 | 800 | 1,000 | 800 |
| Private Offices, Churches, Libraries, Schools | 500–1,000 | 600 | 1,500 | 1,200 | 1,200 | 1,000 |
| General Offices, Deluxe Restaurants, Deluxe Stores, Banks | 1,200–1,500 | 700 | 1,700 | 1,600 | 1,500 | 1,200 |
| Average Stores, Cafeterias | 1,500 | 800 | 2,000 | 1,600 | 1,500 | 1,200 |

**FIGURE 41-2.** This chart shows the recommended air velocities for different applications.

Fortunately, there is a good floor drain system in the room. Bob says, "The dirt is just streaming out of the coil. It was really dirty down in the core. I wonder why it's so dirty when they change the filters so often."

BTU Buddy comments, "This is a post office and they handle a lot of paper. Paper seems to have a fine dust that will eventually stop-up the coil. This type of air filter allows a certain amount of bypass air—air that slips through and carries small particles with it. It would be better for the manager to use a finer filter media. You can talk to him about that. He'll have to change the filters more often, but the coil won't get dirty nearly as quickly. You might also talk to him about putting a filter alarm system on the fan coil, which would measure the air pressure drop through the filter and tell him when the filters should be changed. When a coil starts to become plugged with fine particles, it becomes a super filter. Over time the coil becomes so efficient as a filter that it will stop-up very quickly. This coil has been getting dirty for a long time, and then suddenly it's really stopped-up."

Bob finishes the cleaning job and assembles the system, then starts it back up. After it runs for a few minutes, he checks the liquid line temperature by touching it, and it's cooler than hand temperature. The electric heat is not operating. The system is back to normal.

Bob goes to the manager and tells him what he found and what he has done about it. He then says, "I'll work up an estimate for a filter air-pressure alarm and send it to you so you can submit it to your supervisor. That would be a big help to this system, along with a better filter media."

The manager responds, "Thanks for doing a good job with our system and explaining your recommendations."

While riding away from the job, BTU Buddy comments, "You did a good job back there. Technicians that care enough to do the job right are always in demand."

Bob said, "Thanks. You've been a real inspiration to my work ethics."

# *42*

# A Heat Pump Blowing Cold Air

The first cold weather of the season has arrived, and the dispatcher calls Bob to explain that a new customer has complained that their heat pump is blowing cold air and not heating the house adequately.

Bob thinks he knows what the problem is and thinks it over as he rides to the job. He thinks that the electric heat might not be coming on during the defrost cycle. Some manufacturers use controls that turn the electric heat off during defrost, to improve the unit's seasonal energy efficiency ratio (SEER). When the system is in defrost, the unit is actually in the cooling mode for that time span, and if the customer is standing in the airflow, it will be cool.

Bob arrives at the job with a plan, only to find that he has drawn the wrong conclusion. When he talks to the customer, she explains, "We moved here to the South from New York, and our heating system there blew hot air during the heating cycle. This system seems to keep the house comfortable, but it always blows cool air. For example, I like to sit and sew by the large window in the den, but the air blows on me and makes me cold. The bathroom air outlet blows cold air and it's very uncomfortable when we're just out of the shower."

Bob looks at the arrangement in the sewing area first. The sewing chair is practically sitting on the floor register. As he's thinking over the situation, BTU Buddy says to him, "Bob, remember that a heat pump's outlet air temperature is normally in the range of 100°F when just the compressor is doing the heating. This home has a heat pump that is about 3 years old, and it isn't practical to change the system. The customer needs a heat pump education. Just think and advise, using common sense."

Bob says to the customer, "You're used to a very different type of heating system than the heat pump you have. For example, the maximum air temperature from the heat pump when the auxiliary electric heat is not operating is about 100°F. When that 100°F air is mixed with the room air, before it reaches your skin it could easily be 80° to 85°F. Skin temperature is typically around 90°F. When 85°F air blows on 90°F skin, it seems cool. It will still heat the house to a comfortable 75°F, but blowing on your skin, it can seem uncomfortable. The furnaces that are used up north are usually gas or oil. The air temperature put out by them can easily be 120° to 140°F. When that air reaches your skin, it typically feels warm."

The customer asks, "Isn't body temperature 98.6°F?"

Bob explains, "Yes, but that's the trunk of your body, not the outer extremities like your arms, legs, hands, and feet. They're much cooler than the trunk of your body."

"Well, what can we do to be more comfortable in our house?" she asks.

## MAKING ADJUSTMENTS TO A HEAT PUMP SYSTEM

Bob responds, "Now that you know what the differences are in heating systems, let's see if we can do something in the den first. We can move your chair over a few inches and that will help some. This is a floor register and it's supposed to distribute the air. The air should be directed toward the wall to help warm it so it won't cool the room," Figure 42-1. "I'm going to turn the register around 180 degrees, and point the air away from your chair and more toward the wall. I think that will help a great deal in this room."

"Why don't we just shut that register off?" she asks.

Bob explains, "You could probably do that, but it won't help the efficiency of the whole system. This system depends on all of the airflow to operate at peak efficiency, so shutting off that register could cause the room temperature to drop."

Bob then suggests that they move to the bathroom problem. He looks over the unit and turns the fan to the "on" position to see how much air is coming out of the register. He explains, "The company installed a low sidewall register in here to avoid having to use the tile floor. A sidewall register is not a good choice in a small room, but they're sometimes used anyway," Figure 42-2. "This one seems to have too much airflow. I'll go under the house to see if I can regulate the airflow using the damper to that run."

Bob goes under the house and finds that the installation crew used a 6-inch duct for the bathroom run. It's a much larger duct than necessary for a run of only about 6 feet, but it does have a damper. He turns the damper down to less than 50% of the flow, and starts back upstairs when BTU Buddy says, "Good job, Bob. You're using your head to educate the customer and redirect some airflow. I believe this will get the job done."

**FIGURE 42-1.** The diffuser registers, spreads, and distributes the air over the walls. The walls are where the heat loss or gain will come from.

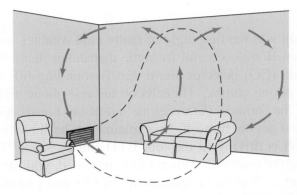

**FIGURE 42-2.** The low sidewall register in this figure shows the air blowing straight out from the wall. If a person is standing or sitting in front of it, they will feel the airflow.

When Bob goes back into the bathroom with the owner, she says, "That really reduced the airflow in here. Do you think it will heat it now?"

Bob says, "Yes, I think it will. Something else that you may want to consider would be to use a small ceiling-mount electric radiant heater—you've probably seen them in motel bathrooms. That would give you that warm feeling when just getting out of the shower. It would have to be installed by an electrician."

He then resets the fan switch on the thermostat to the "Auto" position and asks, "Are there any more areas that are uncomfortable?"

She says, "No, I think that's it. Now I have an idea of how to deal with a different type of heating system, and what to look for. Thank you."

As they're leaving, BTU Buddy says, "Good job, Bob. Getting the customer to understand the system was a real help in this case. Using your good judgment definitely helped on this job. You're getting better and better at customer relations and service work. Customer relations is a big part of having a happy customer."

# Carbon Monoxide Problem with Gas Furnace

Last night was the first night of really cold weather, and it's a cold morning, too, when Bob gets a call from the dispatcher that a customer has a carbon monoxide (CO) detector alarm that is sounding off. Bob asks for the customer's phone number. He calls on his cell phone and tells the customer to shut off the furnace, get everyone out of the house, open the front and back doors, and get out himself. He explains that he'll be there within 30 minutes.

As Bob is driving to the job, BTU Buddy says, "That was good advice you gave the customer. When CO is involved, you can't be too cautious."

Bob says, "I've been reading up on CO and I know that it's better to be safe than sorry."

Bob arrives at the house and sees that the front door is open and the family is sitting outside in the car. The homeowner then steps out of the car to talk to Bob and says, "This has put us in an uncomfortable position, waiting in the car. Was it necessary?"

Bob explains, "Carbon monoxide is very dangerous. It isn't like breathing some smoke where you can step outside and clear your lungs. CO accumulates in the body and does not just clear out with fresh air. It's vital to not let it accumulate. Tell me what happened."

The owner tells Bob, "I bought two CO detectors at a local hardware store last week," Figure 43-1. "One was for our house and the other was for our daughter and her family at their house. I plugged ours in last week and it has been fine—no alarms. Last night it was cold, and I built a fire in the den fireplace. Later in the night, the CO alarm went off. I got up and looked at it, and decided that it must be defective, so I unpacked the second alarm and put a battery in it and plugged it in. I went back to sleep and about 30 minutes later it went off. I just figured they must both be bad, so I packed them back up. When I got up this morning, I decided to try one of them again

**FIGURE 43-1.** A carbon monoxide detector that plugs into a 115V wall outlet.

and it went off while I was getting dressed. I decided to call your company to do a checkup."

Bob says, "I'm glad you called us. It's best to be safe about this. I have an instrument to detect CO in the house, but there's no sense in testing now, with the doors open. I suspect that the fire in the fireplace plays a part in this. I'll inspect everything and find the problem before I leave. Is there still a fire in the fireplace?"

The owner says, "Yes, I built it up before I went to bed and there were plenty of coals left this morning. I put several logs on the fire when I got up. It should still be burning."

Bob goes into the house and shuts the back and front doors. Then he goes to the furnace room, which is shared with the washer and dryer. The furnace is a natural draft 80% efficiency furnace. Bob puts his hand near the draft diverter and feels fresh air coming down the flue. It's just as he thought. The fireplace flue is very hot from the big fire, and pulling a vacuum on the whole house, causing the gas furnace to back draft, Figure 43-2. Now, the question is, where did the carbon monoxide come from? A gas furnace does not normally give off carbon monoxide.

Bob removes the front panel from the furnace and can easily see what the problem is. The furnace has slotted-type burners with adjustable air shutters, as shown in Figure 43-3. These air shutters have a lot of lint in them. They've been burning with a yellow air-starved flame, and have built up quite a bit of soot in the furnace. This is going to be a furnace-cleaning job.

Bob goes back outside to the homeowner and tells him that the family can come back in, as the house is completely aired out. He then explains what happened. He also explains that, when a fire is burning in the fireplace, it needs to get plenty of air by having a window in the room slightly open. Bob asks the owner to go to the furnace room with him and with a candle flame, shows him what was happening.

Then he goes to the den where the fireplace is located, and opens an outside window about 3 inches. They go back to the furnace room and the candle shows that air is moving up the vent, even without the furnace running. This is the way it should be. The homeowner says, "This is the first time a service technician has ever showed me how a system works. Thanks."

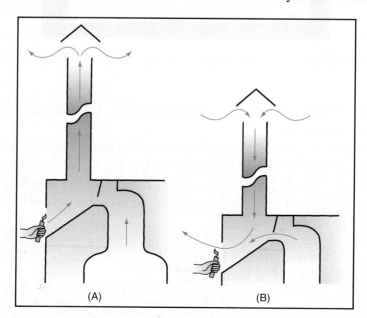

(A)                                    (B)

**FIGURE 43-2.** (A) Shows how a furnace flue with a natural draft should vent. (B) Shows a natural draft furnace that is down drafting. The candle held in front of the draft diverter shows which way the air is moving.

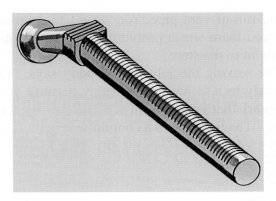

**FIGURE 43-3.** This is a cast iron burner that has an air shutter to adjust the air. These burners can become stopped-up with lint over time.

Bob goes back to the truck where BTU Buddy says, "Bob, that was a great job of finding the problem and handling it. Now, for cleaning the furnace. All of that soot will have to be removed. Do you know what to do?"

"I do," says Bob, and he goes to work.

There is a lot of soot in the furnace heat exchanger so Bob removes the draft diverter and uses long brushes and a shop vacuum to clean the top end of the furnace. He removes the burners and takes them outside, and uses compressed air to blow them out. He cleans the area where the burner slots are located, again using the brushes and shop vacuum. When the furnace is clean and the burners are replaced, Bob starts the furnace. The flames are orange at first because of all of the dust in the air. When the furnace begins to get hot, the flames become blue, as they should be. Bob replaces the furnace door and rechecks the flue with a candle, and it's drafting as it should.

Bob returns to the homeowner and says, "Fireplaces are nice, but they can create problems if they aren't operated correctly. They're not very efficient for heating unless the air for combustion can be pulled from outdoors, without coming through the room. The most heat you get from a fireplace is the radiant heat from the glowing fire."

The homeowner says, "I didn't know any of these things about operating an open fireplace. Now that you've explained it, I understand. A fire is mainly for decoration unless managed correctly."

"That's correct," says Bob. "Your system is in good shape now. Having the dryer in the room with the furnace is not a great combination, but that's about all you can do in this case. It's a good idea to make sure the dryer vent is always clean. At least this one goes through the wall and doesn't have a long run of vent pipe. Your new CO detectors really did their job. They told you there was a problem. I'm glad you called us instead of just returning them to the store."

As they're leaving the job, BTU Buddy says, "I can see that you'll be a BTU Buddy to a technician someday because you really care about the quality of work that you do."

Bob says, "Thanks—that's a compliment."

# Frozen Heat Pump Outdoor Coil

It's another cold day after a week of below freezing temperatures, when Bob gets a call from the dispatcher telling him that a customer has a heat pump outdoor coil that's frozen solid.

Bob arrives at the house and talks to the homeowner, who says, "The auxiliary heat light on the thermostat has been on for most of 3 days, and the outdoor unit has a huge sheet of ice on it. I noticed that the neighbor's heat pump doesn't have ice on it."

Bob says, "That was good thinking, to look and compare your unit to the neighbor's unit."

Bob goes outside and notices that the compressor is running, but the fan is not. The coil has inches of ice on it. He goes to the truck and gets his tool pouch and a meter to check out the fan. He removes the cover to the compressor compartment to get to the electrical leads. He starts to shut off the disconnect when BTU Buddy appears and says, "Why don't you leave the unit on for the initial check. If the problem is in the defrost circuit, it may correct itself when you shut it off, and then you won't be able to determine what it was. The first thing to check is whether you have power going to the fan."

Bob carefully checks the power to the fan leads. There is power, but the fan is not running. He wonders how that can be, when BTU Buddy says, "It's either the motor or the capacitor. If the capacitor is defective, the motor will stop on its internal overload," Figure 44-1. "Or it may be a motor that has inherent protection. Inherent protection is for small motors that won't get excessively hot if they stall. They can just stay in a locked rotor condition and won't get hot enough to actually burn."

Bob asks, "What do we do next?"

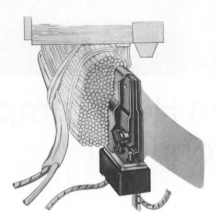

**FIGURE 44-1.** This overload protector is actually wound into the motor winding.

BTU Buddy says, "Turn the unit off at the disconnect and let's investigate one thing at a time."

Bob turns off the disconnect and removes the panel to access the fan motor and says, "This motor is hot, so it must have a winding because it has been drawing current to get hot and it must be off because of heat or overload."

BTU Buddy suggests, "Short out that fan capacitor before you start. It may have a charge stored in it. Remember that the manufacturer wants you to use a resistor to short the capacitor to prevent excess load on it."

Bob uses a resistor and shorts across the capacitor. There is no spark and he asks BTU Buddy, "Why?"

BTU Buddy explains, "The capacitor will often discharge through the motor winding or just may not be charged. It's much safer to discharge it. You could hurt yourself backing up if it discharges through you. You could also damage your meter trying to check a charged capacitor. Capacitors can really have a jolt in them. I've seen technicians use a screwdriver to short across them and the arc would leave a mark on the screwdriver shaft."

Bob labels the motor leads and removes them; then he uses his ohmmeter to check the motor. All of the windings have resistance. He moves his meter scale to R × 10,000 to check if there's any circuit to ground. There is not. The motor is electrically sound. He then uses a capacitor tester, as in Figure 44-2, and checks the capacitor. It is defective. He goes to his truck for a fan capacitor and replaces it. He's ready to start the unit. He asks, "What are we going to do about all of that ice? It will take a long time and many defrost cycles to get rid of it."

**FIGURE 44-2**. This digital tester tells the actual capacitance of the capacitor. *(Courtesy Davis Instruments)*

BTU Buddy says, "Leave the unit power off and get a water hose. We'll use that water to melt the ice. The water will be at about 45°F and will quickly defrost it. It's not solid ice; it has a lot of air in it. If you try to defrost it using the defrost cycle, it will quickly melt the ice close to the coil. The outside ice will then act as a shell and restrict the airflow. It's best to get the coil clean and then start it."

Once the coil is defrosted, Bob starts the unit, and all seems well. The gas line becomes hot quickly, a sign that the charge is good. He lets the unit run for a little while and goes inside to check the air temperature at the registers. It's hot. The auxiliary heat is still working with the heat pump.

Bob puts all the panels back in place and talks to the homeowner. She says, "Thanks for the quick response. You've been working out in the cold for a while. Let me fix you a hot cup of coffee to take with you to the next job."

She fixes the coffee for him and as he's driving off, BTU Buddy says, "Wish I were real and could have some of that coffee. It's really cold out there. That was nice of her to do that for you. It's a sign that you did a job that she approved of."

# Burned Transformer Due to Shorted Coil in Relay

Bob receives a call from the dispatcher about a customer that has a system that's not heating. The system is a heat pump with an electric furnace, and the customer explained that nothing about the system was working this morning as she was trying to get the family started on their day.

Bob arrives and talks to the customer to see what he can learn. She says that when she got up, the house was cold, and when she went to the thermostat and pushed the lever up for heat, the furnace didn't start as usual. They go to the thermostat and Bob pushes the lever to "fan on," but the fan doesn't start up. He tells the customer that he'll get started on the job and find the problem.

The indoor unit has a 230 V power supply so he checks the voltage going to the unit, 235 V. That's as it should be. Then he checks the power to the low voltage circuit, which is the load side of the 24 V transformer. The voltage is 0 V. The transformer has voltage going in, but no voltage coming out. This is simple enough, he thinks.

He turns off the power and changes the transformer, attributing the problem to random transformer failure. Little does he know that there are other problems.

He's ready to start the unit when BTU Buddy appears and asks, "What do you think caused that transformer to fail?"

Bob says, "I think it just randomly failed."

BTU Buddy says, "You could be right, but doesn't it make you wonder why it worked for years and then just failed?"

Bob says, "No, it just looked to me like it failed."

BTU Buddy then says, "A professional technician always looks for why. Otherwise you just become a 'parts changer.' Oftentimes that will come back to bite you."

"What do you think I should do?" Bob asks.

BTU Buddy responds, "Transformers are notorious for being reliable. When one has lasted for years, it may have been subject to some stress that caused it to fail. Let's look for that stress to find out if that's what happened."

Bob asks, "How do we go about that?"

"Use a 10-wrap coil of wire to check the amperage in the low-voltage circuit when you start the furnace up," BTU Buddy explains, Figure 45-1. "This system has a 40 volt-amp transformer. The maximum amperage that the low voltage of the transformer can output without overheating is 1.66 amps (40 volt/amps divided by 24 volts equals 1.66 amps). This amperage is so low that your ammeter may not measure it. The 10-wrap of wire will amplify the amperage 10 times, so the amperage should not exceed 16.6 (10 times 1.6 equals 16.6 amps). Set the ammeter scale to 50 amps."

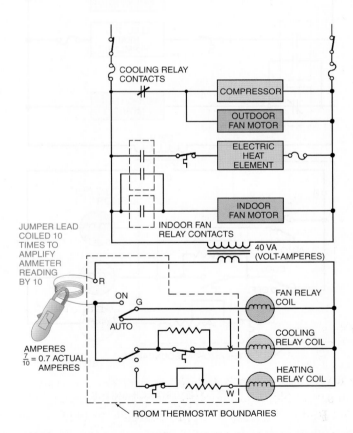

**FIGURE 45-1.** A clamp-on ammeter can be used to measure very low amperages by wrapping the wire to be checked around the ammeter's jaws to amplify the reading. Ten wraps will amplify 10 times. The true reading can be arrived at by dividing the meter reading by 10.

Bob gets the system ready to start with a 10-wrap of thermostat wire in the circuit. With the power off at the air handler, he then sets the thermostat to call for heat and returns to the furnace to start the system. When he turns the system on, using the disconnect, the amperage in the low-voltage circuit goes up to 25 amps, or an actual 2.5 amps in the circuit. Bob shuts the

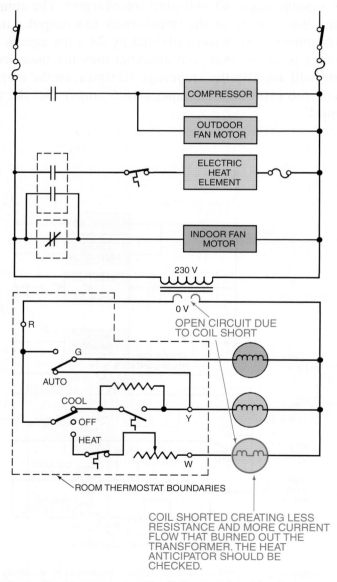

**FIGURE 45-2.** The heat relay in this circuit has a shorted coil. Some of the coil's wires are touching internally, causing the resistance to be less than normal.

system back off, scratches his head, and says, "Now what? I can see that the amperage is too high, but where is the problem?"

BTU Buddy says, "There may be some wires touching, or there may be a 'shorted' coil. A shorted coil happens when some of the coil turns are touching together, causing it to draw too much amperage. As a result, the resistance in the coil is less than it should be. Since there are only two coils in this circuit—the fan and the electric heat coil—let's check them for resistance. You have some low-voltage coils in the truck that are new. Let's compare them with the ones in the system and see if there's a big difference."

Bob brings in a typical 24 V heat relay and a fan relay for comparison. When he compares the heat relay in the system with the heat relay from his truck, it shows much less resistance.

"That heat relay is shorted," BTU Buddy says, Figure 45-2. "You almost changed that transformer and left, without noticing that it was drawing too much amperage. It would have quickly burned the new transformer, and you would have been back to find the real problem. Some manufacturers put inline fuses in the system to prevent the transformer from failing. This one didn't have that type of fuse, so the transformer overheated and failed."

Bob says, "I'm glad you insisted that I do the right thing. I really hate callbacks. The company doesn't like them either."

BTU Buddy agrees, "When a technician gets called back to a job, the company has to pay the technician to do the job again, when it should have been done correctly the first time. The company cannot collect for these callbacks, so management frowns on them. An important part of being a professional is to do the best you can the first time, to eliminate a second call at no charge to the customer."

Bob says as they are driving away, "It's not easy to be professional all the time."

"That's why there are fewer truly professional technicians," BTU Buddy says. "Too many people will settle for taking the easy way out."

# Handling a Stopped-Up Oil Filter

Bob gets a call from the dispatcher that a residential customer out in the country has an oil furnace that has to be reset to get it to run. Bob tells the dispatcher to call the customer back and tell them not to reset the furnace again until he gets there, in about an hour.

It's a very cold day, so Bob gets there as soon as possible. When he arrives, the customer shows him where the furnace is located down in the basement. As Bob suggested, it has been left off. He knows to check the furnace for excess oil before he resets the primary control, and also not to start up a furnace that has oil standing in the combustion chamber. Since the furnace is not hot, he rolls up his sleeve and puts his hand into the combustion chamber and discovers that it's dry. An examination with his flashlight shows a lot of soot in the area of the combustion chamber and heat exchanger.

Bob activates the primary control reset. When the furnace fire starts, he looks through the inspection door at the flame, which is unsteady and doesn't have the proper color—orange with yellow tips. It's also lazy and smoky. He asks the owner when the burner was last serviced. The owner says it was over a year ago, so Bob thinks the burner just needs a good service call.

Bob shuts the furnace down and turns off the disconnect. He gets his oil burner service tools from the truck and removes the burner, setting it on the floor. He uses a nozzle wrench, as in Figure 46-1, to change the nozzle with an exact replacement. He cleans the insulators with solvent and sets the electrodes with his electrode gauge. Then he carefully removes the cad (cadmium sulfide) cell.

About this time BTU Buddy appears and makes a suggestion, "Be careful not to touch the pins that plug the cell into the socket with your fingers.

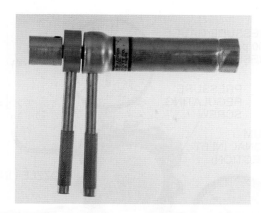

**FIGURE 46-1.** This special nozzle wrench should always be used to remove oil burner nozzles. It holds the nozzle holder and enables the technician to turn the nozzle. Using adjustable wrenches can warp the electrodes.

The current in that circuit is very low and any resistance, such as oil from your hands, may cause a problem."

Bob says, "I didn't realize that could be a critical point."

BTU Buddy responds, "Yes, that connection must be very clean. I see that the cad cell is dirty. What do you think caused that?"

"I'd suspect the burner is burning with a dirty, smoky fire," Bob says.

BTU Buddy says, "I think you're right. That fire was really bad. I hope changing the nozzle will help. Was the strainer behind the nozzle dirty?"

"It was really dark, like it had been filtering out something," Bob replies.

Bob puts everything back together and is ready to fire the furnace. It fires right up, and as Bob looks at the flame he comments, "The flame still doesn't look good. I wonder what's going on?"

BTU Buddy says, "I think you should check the oil pressure. It may not be correct."

Bob installs gauges on the pump, Figure 46-2, and starts the furnace. "The pressure on the pump inlet is below atmosphere, and the outlet pressure is low," he says.

BTU Buddy says, "You didn't change the inline oil filter," Figure 46-3.

Bob says, "I didn't think to do that, but I'll do it now."

He turns off the oil at the tank and places a pan under the filter housing to catch any oil that may spill. He removes the filter cartridge and sees that it's really dirty, with a lot of rust and scale in it. He changes the cartridge, and refills the housing with fresh oil to prevent an air pocket from being in the line on startup.

While Bob is changing the filter, BTU Buddy says, "The tank is above ground and just outside the basement. The pump suction shouldn't be in

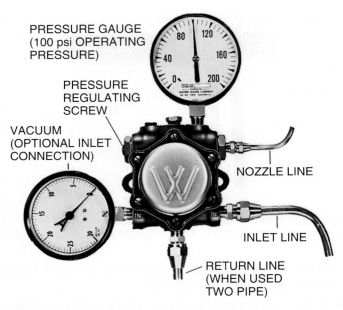

PRESSURE GAUGE
(100 psi OPERATING
PRESSURE)

PRESSURE
REGULATING
SCREW

VACUUM
(OPTIONAL INLET
CONNECTION)

NOZZLE LINE

INLET LINE

RETURN LINE
(WHEN USED
TWO PIPE)

**FIGURE 46-2.** Test gauges fastened to an oil pump. *(Courtesy Webster Electric Company)*

(A)                          (B)

**FIGURE 46-3.** Inline oil filter cartridge (A) and housing (B).

a vacuum because the oil level is above the burner. The inside of that tank must be getting rusty, as scale is forming. I'll bet the owners have been trying to save money and not keeping the tank full during the summer. In that situation the tank will take in cool, moist air at night, then it expels part of that air the next day when it warms up. It's best to keep the tank as full as practical, particularly during the summer. Many owners let it run low and don't buy again until fall, but that's not good practice.

"You can change the filter and it may or may not get dirty again. The first season of burning may have removed enough dirt from the flowing oil."

"What do I tell the owner?" Bob asks.

BTU Buddy says, "The truth. Tell her that you may need to come back and change the filter again, and tell her why. In fact, it would be good to set up another filter change in about a week, so you can inspect the filter."

Bob restarts the furnace. The gauges are now reading about 1 psi entering and 100 psi leaving the pump. Bob says, "This looks good to me."

He removes his gauges and puts the furnace back in operation. He checks the air filter and finds that it needs changing, so he changes it and oils the motors. He's ready to leave when the owner comes down to the basement.

"It's good to feel heat coming out of the system again," she says. "Thanks."

Bob explains the situation to her and she tells him that, sure enough, they let the tank sit nearly empty all summer, but that they'll follow his advice for next year. Bob sets up a time to come back and check the oil filter.

As they're driving away, BTU Buddy says, "Good job. That's another happy customer. One by one, your company is building a customer base that believes in you and has confidence in your abilities."

Bob says, "There sure is a lot to being a service technician."

BTU Buddy says, "There's even more to being a competent, professional technician, and you are doing it."

# A Smoking Oil Furnace

Crisp, fall weather has moved in, and people are starting to turn on their furnaces in the morning. Bob gets a call from the dispatcher saying that a customer was smelling smoke and oil fumes from her oil furnace in the basement. Bob calls the customer and tells her to shut the furnace off until he can get there to check it out.

He arrives and talks to the customer, who leads him to the upflow oil furnace in the basement. He can still smell the oil fumes. He shuts off the switch at the furnace, then goes to the room thermostat and sets it to call for heat. This allows him to operate the furnace from the basement switch when he wants it to run.

Bob turns on the switch and the furnace starts up normally. Within a few seconds, oil smoke begins to seep from around the burner flange and the inspection door to the burner. He shuts it off before it becomes too hot.

Bob thinks to himself that this is a classic case of needing a burner tune-up. He brings his oil burner tools in from the truck, removes the burner, and sets it on the floor so he can change the nozzle. As a precaution, he checks the nameplate and confirms that the 0.75 nozzle was the correct size. But it looks old and stained, so he knows he's on the right track. After changing the nozzle, he changes the oil filter, carefully placing it in a small pan that he carries just for containing oil spills. Then he fills the filter cartridge housing with fresh oil from a gallon that he carries for this purpose. He knows that filling the cartridge with fresh oil will keep air from entering the system.

This is a two-pipe system so Bob fastens a gauge on the inlet to the pump and the outlet to the pump, to verify the correct oil pressure. He's now ready for start-up.

He turns the furnace on while looking at the gauges. They read 100 psig on the outlet and 3 inches of vacuum on the inlet. He's thinking about what a great job he's doing, when smoke suddenly begins to form again in the same places. Now his confidence has gone out the window. He knows that all oil furnaces need to have a routine nozzle and filter change. He made sure the burner was tight against the flange, yet smoke is still filling the area. BTU Buddy appears and asks, "What's the trouble, Bob?"

Bob goes over all that he's done.

BTU Buddy says, "The nozzle and filter definitely needed changing—that was the right thing to do. But you aren't finished with this call. There are some other routine things that you owe the customer. Start with checking the draft over the fire. Then we'll move on to a smoke test and a $CO_2$ test for efficiency. I see that you have all of the instruments you need."

Bob finds a level spot for his draft gauge and sets it to 0, Figure 47-1. Then he places the probe in the furnace access door and starts the furnace. In just a few seconds, the draft gauge shows a positive pressure. Bob says, "This system is showing a positive pressure. It's not drafting."

BTU Buddy asks, "What do you think the problem might be, and what are you going to do first?"

Bob says, "I'm going to loosen the flue connector to the chimney and look in there to see what I can see."

He loosens the connector and sees that it's clean; this problem hasn't been going on for long. He shines a light into the furnace to find that it's not very dirty—no soot accumulation—verifying that the problem is a new one. He then shines the light into the chimney and notices some straw at the bottom. He uses a mirror to look up the chimney. He sees light, but not much.

Bob puts a ladder against the side of the house and goes up on the roof. He shines his light down the chimney and sees the problem. There is a bird's nest partway down that is partially blocking the chimney.

**FIGURE 47-1.** This figure shows a draft gauge that is being hand-held. It actually should be placed on a level solid surface, not hand-held.

Bob goes to his truck to figure out how to remove the nest. He doesn't really want to push it to the bottom, so he fastens a piece of stiff wire to his flue cleaning rod and bends it to look like a fish hook. He goes back to the roof and hooks the nest, pulling it out. Now he has something that he can show to the homeowner.

Bob goes back to the furnace and starts it up again. It starts fine and doesn't smoke. The draft gauge shows 0.02 inches of draft, which is a good draft for over the fire, Figure 47-1.

BTU Buddy then says, "It's time to perform the rest of the testing—a smoke test and a $CO_2$ test."

Bob does a smoke test, which is negative, Figure 47-2. The $CO_2$ test shows 13% $CO_2$, which is very good, Figure 47-3. The flue temperature is 510°F and the temperature of the basement is 65°F, a net flue temperature of 445°F, and the efficiency is over 75%. All is well.

Bob shows the nest to the homeowner and explains to her that he did a complete fall tune-up, and on his report he shows the readings that he took. He advises her to have some sort of protection installed on top of the chimney to prevent more bird's nests; they're likely to return to the same place

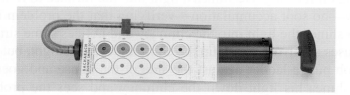

**FIGURE 47-2.** This is a smoke test pump and samples of smoke test colors.

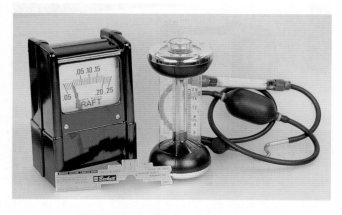

**FIGURE 47-3.** This is a $CO_2$ analyzer.

next year. She tells Bob that she knows who to call to get that set up, and will do it before spring.

While Bob is driving away, BTU Buddy says, "You tried to take a short-cut back there and proved to yourself that shortcuts don't pay."

Bob says, "You're right. I assumed that a tune-up would solve everything."

BTU Buddy says, "It's always best to run an efficiency test and a smoke test. They take a little more time, but now you can be sure the system is in good shape, and you verified it in a written report to the customer. That builds customer confidence in you and your company."

"Again, thanks for helping me become a professional," Bob says.

# Grounded Electric Heat Coil Causing Fan to Run

Bob receives a call from the dispatcher about a customer whose heat pump indoor fan motor is coming on from time to time for no apparent reason. It's happening even when the thermostat is in the "off" position.

Bob goes over to the house to talk to the customer, who says, "The thermostat is set in the 'off' position today because it's so mild that we don't need either heat or air conditioning. But the indoor fan starts up from time to time and runs for a few minutes, and then shuts right back off, for no apparent reason."

Bob says, "I can't imagine what that could mean. I'll check it out."

Bob goes to the garage where the indoor unit is located, and starts to look around. While he's standing there, the fan starts up and runs for about two minutes and then shuts off. Bob scratches his head and thinks. He removes the fan wire from the thermostat to the terminal board so there's no chance for a circuit to be reaching the fan start relay. Then he waits to see if this solves the problem, when the fan starts up again. Now he's really confused.

Just then, BTU Buddy makes an appearance and says, "You look confused. Tell me about it."

Bob explains to BTU Buddy what has been happening and BTU Buddy says, "Tell me what can stop and start the fan on this system."

Bob says, "The fan relay starts the fan upon a call for heat. The fan must be running while the compressor is running. That's all I know."

"Look closely at the components on the electric furnace and look at the fan circuit," BTU Buddy says.

Bob looks over the furnace fan circuit and says, "It looks like there's a temperature sensor that can start the fan. I wonder what that's for?"

BTU Buddy replies, "If the electric heat were to start up for any reason without the heat pump running, for example in emergency heat, what would start the fan?"

Bob says, "Well, I guess the temperature-activated fan switch would."

BTU Buddy says, "What would keep the fan running during electric heat operation, to cool the furnace down at the end of the cycle? Now feel the furnace and see if you can detect any heat on the surface."

Bob runs his hand over the furnace and said, "Yes, it's warm, not hot."

BTU Buddy says, "Okay, get your ammeter and see if any of the heater circuits are drawing current."

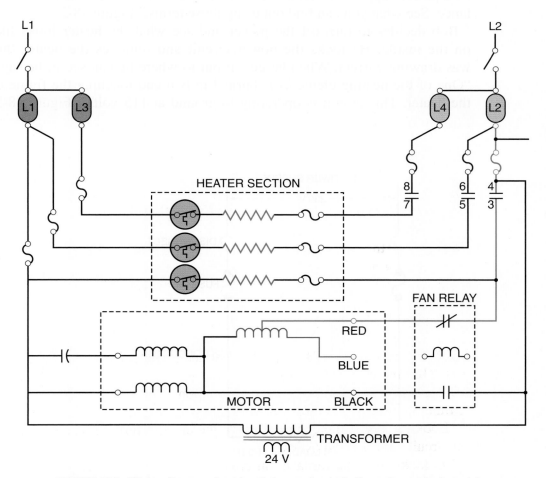

**FIGURE 48-1.** Wiring diagram of an electric heater circuit similar to the one in the article.

Bob uses his ammeter and checks all of the circuits. The furnace has three stages of electric heat and Bob says, "One of the circuits is drawing 10 amps; it should draw about 15 amps."

"Put your ammeter on the other side of that heater and see what you get," says BTU Buddy.

Bob checks the amperage on the other side of the heater and gets nothing. He says, "Boy, am I confused now."

BTU Buddy says, "Check the voltage to the heater. It should be 230 volts."

Bob checks the voltage, Figure 48-1, and says, "There's no voltage, yet it's drawing current."

BTU Buddy says, "Remember the basic law of current flow: there must be a power supply—something to conduct the power—and a load or resistance. See what you can find out using those terms," Figure 48-2.

Bob decides to turn off the power and see what the heater looks like on the inside. He locks the power circuit and removes the heater that was drawing current. When he gets it out to where he can see it, he says, "One of the heating elements is burned in two and touching the frame of the heater. This circuit is operating to ground at 115 volts," Figure 48-3.

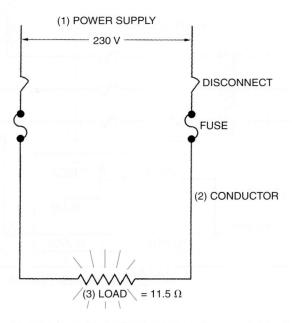

**FIGURE 48-2.** This diagram shows the three conditions for current flow: a power source, a conductor, and a load or resistance.

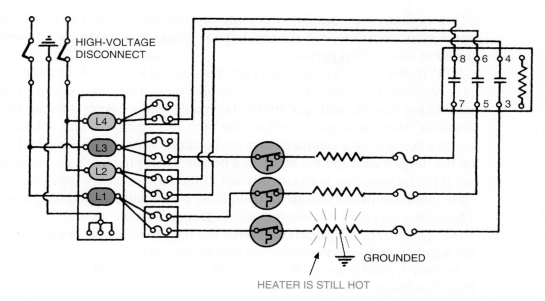

**FIGURE 48-3.** This heating element is grounded to the frame and operating off of 115 V. Note, if the frame of the furnace is not grounded it would not work, and the furnace would be electrically hot.

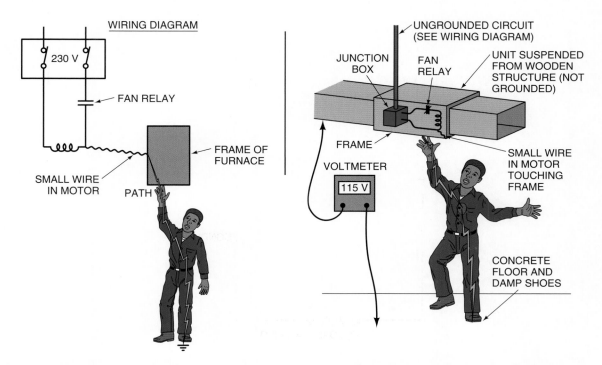

**FIGURE 48-4.** This electric motor has a ground circuit in it and the furnace casing is not grounded. The technician becomes the path to ground. This is very dangerous.

"That explains why the fan is coming on. It's enough heat to cause the temperature sensor to start the fan, but not enough to keep the fan running for more than a short period of time."

BTU Buddy says, "Good work, Bob. You figured it out without too much of a problem. The heater to ground explains it all. You could see from the installation that this unit was installed correctly, and we were standing on a dry floor when we started checking the unit. But anytime a unit is located under a house or in a wet basement, it's good practice to put one meter lead on the furnace case and the other on a nearby water pipe that's grounded to see if the furnace case is electrically hot. If the furnace case weren't electrically grounded in this situation, it would be electrically hot and dangerous. The technician could become the path to ground," Figure 48-4.

Bob explains to the homeowner what has happened, and then goes off to the supply house for a new heater, which he then installs.

As they're riding away, Bob says, "Boy, I never would have expected that. This business is full of surprises."

BTU Buddy says, "You accumulate these experiences as you go along. I'm glad we were able to find the problem."

# Air in a Hot Water Heating System

It's a 30°F day, and the dispatcher sends Bob to a motel that's five stories high, because of a complaint that the top floor has no heat. People are having to move to the lower floors to stay comfortable, creating an expensive problem for the motel.

When Bob arrives, he asks the manager to explain what he knows about the problem. The manager says, "Late last night the front desk began to get complaints about it being cold on the top floor. By morning, it was obvious that we had a problem."

Bob goes to the equipment room in the basement where the boiler is located for a look around. He discovers that there is a hot water boiler with one hot water pump that services the entire building. The boiler is hot, the pump is running, and the sound of water circulating is noticeable. Everything seems okay at the boiler.

He then goes up to the fifth floor, where the problem is, and checks the fan coil unit in one of the guest rooms, Figure 49-1. The fan is running, but cool air is coming out of it. The unit has a thermal operated valve to control the hot water to the fan coil. Bob thinks that it must not be opening, Figure 49-2, so he takes a voltage reading at the valve. It read 24 volts and the indicator on the valve shows it to be open. He goes to another room and finds the same thing. He's beginning to get confused when BTU Buddy shows up and asks, "What's the problem, Bob?"

Bob goes over what he has checked so far, and BTU Buddy says, "It sounds like you're on the right track. You said you went to the boiler room and the boiler was hot. Did you check the entering and leaving temperatures?"

Bob says, "No, but that would be a good thing to know."

They go down to the boiler room and find that the leaving temperature at the boiler is 190°F and the return water temperature is 160°F.

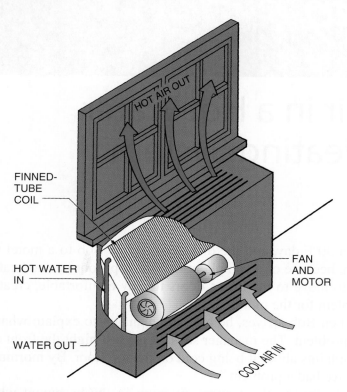

FINNED-
TUBE
COIL

HOT AIR OUT

HOT WATER
IN

FAN
AND
MOTOR

WATER OUT

COOL AIR IN

**FIGURE 49-1.** Typical fan coil unit that may have a chilled water circuit for cooling and a hot water circuit for heating.

Bob says, "There seems to be plenty of heat, but it's not getting to the fifth floor."

BTU Buddy suggests, "I suspect there's air in the system. If air gets into the water side of the system, it will naturally go to the top and the water won't circulate. Let's go to the top floor and see what we can find."

They go to the fan coil unit that Bob checked before and BTU Buddy says, "Use your screwdriver as a 'poor man's stethoscope' to see what you can hear in the piping," Figure 49-3. "If there's any water circulating, you should be able to hear it."

Bob listens and says, "I don't hear anything moving in the pipe."

BTU Buddy says, "Let's find the high point in the piping and see if we can find an air bleed valve. The piping normally runs to the top floor and then branches out to the various circuits, and there should be a bleed valve on the supply riser and the return pipe. It would save time if we could see the blueprints of the building. Let's go to the office and see if they have them."

**FIGURE 49-2.** This is a thermal action zone valve. It is not a snap acting valve but has a 24 V heat motor inside. It is very quiet in its action as opposed to a solenoid snap acting valve. *(Courtesy Taco, Inc.)*

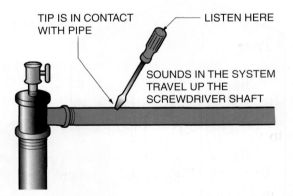

**FIGURE 49-3.** We call this a poor man's stethoscope. It is used to listen for vibration or water flow.

They speak to the manager and he gets out the prints. They find that the bleed valve is behind an access panel in the ceiling on the fifth floor.

BTU Buddy says, "Get a length of flexible plastic tubing and a stepladder from your truck, and a bucket to catch the bleed water to keep it off of the carpet."

When Bob returns, they find the removable panel and the bleed valve, and make the connection.

They connect the bleed line to the valve and let the end down into the bucket, ready to bleed the air, Figure 49-4.

Bob opens the bleed valve and nothing but air comes out. BTU Buddy says, "The upper portion of this system is air logged. It will take a few minutes for all of the air to bleed out."

After several minutes, water begins to trickle into the bucket. The water is rusty and dirty, and Bob says, "Boy, the tube and bucket were a good idea. Rusty water on this beige carpet would look awful, motel management would not be happy."

BTU Buddy says, "Your company might not be invited back if that happened. Rust is very difficult to remove."

After a few minutes, the water becomes hot. BTU Buddy then says, "Now bleed the return line to make sure that all of the air is out of the system. Then leave the bucket, ladder, and tube here. We have more investigating to do, but first let's see if we have heat yet."

They go back to the two rooms they had previously checked and find hot air coming out of the fan coil units. Bob says, "Well, thanks to you we're through, and all is well."

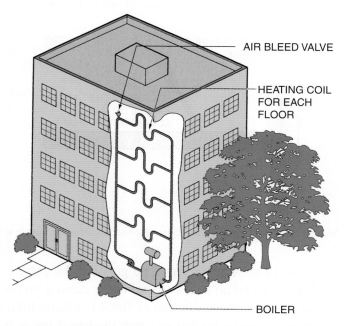

**FIGURE 49-4.** This system includes a bleed valve at its high point. The system is a closed loop of water pipe, and the air will go to the highest point in the system.

BTU Buddy says, "Most technicians would leave the job now, but I think we should find the cause of the air in the system. If we walk out now, it's likely to occur again. You have to go on another call, so let's meet back here tomorrow and solve the problem."

Bob asks, "Will the system run like it should until tomorrow?"

BTU Buddy says, "Yes, this air accumulation most likely happened over a period of time. This is a closed-loop piping system, so air will be slow to accumulate. Get your ladder and leave the bucket and tube above the ceiling, and put the access panel back until tomorrow."

Bob explains the problem to the manager and tells him that he'll be back in the morning.

*Look for Part 2, "Service Call 50: How Air Gets into a Hot Water Heating System," in the next chapter.*

# How Air Gets into a Hot Water Heating System

Bob and BTU Buddy have returned to finish a job from the previous day. There had been so much air in the hot water heating system of a five-story motel, that the top floor had no heat. In the initial call, they purged the air from the system by bleeding it at its highest point. That got the system working properly again. Now they want to determine how the air got into the closed-loop piping system.

Bob asks, "What do you think the cause could be?"

BTU Buddy says, "Two things come to mind. If the system has a leak, and water is being lost, the makeup water could cause air to build up in the system. City water is used for the makeup water and it contains air. You probably have noticed that if you leave a glass of water out on the table, that air bubbles appear in it," Figure 50-1. "If enough city water is allowed into the system from leaks, air will accumulate at the high point in the system. Another possible cause is that the hot water pump is sucking in air

AIR BUBBLES ON SIDE
OF GLASS COME FROM
CITY WATER

**FIGURE 50-1.** This standing glass of water from the city water supply shows that there is air in the water. To see this, you can fill a glass with water and watch what happens after it stands for a while.

around the pump shaft seal," Figure 50-2. "We should do some checking around for leaks and test the pump inlet to see if it's running in a vacuum."

"How are we going to check the system for water leaks?" Bob asks.

BTU Buddy says, "If there are water leaks at the individual units, water will be dripping from the condensate drain lines. Notice that all of the units are on an outside wall. You can see the drain lines coming out the wall, and they drip outside into the flowers or on the ground. Let's walk around the motel and see if any of them are dripping."

They walk around the building and find two lines that are dripping water. They go to those rooms and remove the covers from the units, finding small leaks that can be tightened up and stopped.

Bob says, "I wonder how long those have been leaking?"

BTU Buddy replies, "Probably a long time. For every bit of water that leaks out, water must be made back up with city water. As we talked about earlier, city water has air in it. I think we've solved the leak problems from the fan coil units. Let's go to the boiler room and see if we find any leaks there."

They go to the boiler room and find a leak at a valve packing on one of the pipes. They tighten up the valve-packing gland, and the leak stops, Figure 50-3.

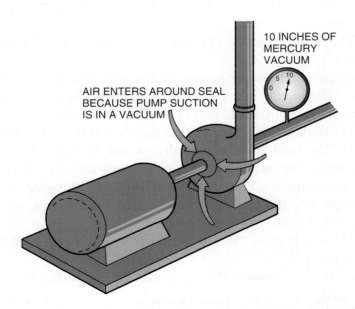

10 INCHES OF MERCURY VACUUM

AIR ENTERS AROUND SEAL BECAUSE PUMP SUCTION IS IN A VACUUM

**FIGURE 50-2.** This pump seal is running in a vacuum. If the seal leaks while it is running, air will be sucked in.

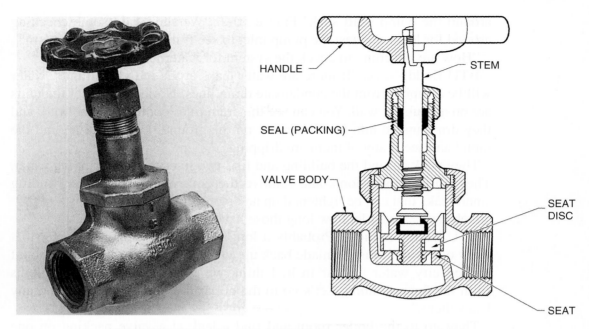

**FIGURE 50-3.** The packing gland can be tightened to prevent it from leaking.

Bob says, "I think we have all of the obvious leaks repaired. If there were more, they'd likely show because they'd be in the piping in the walls and ceilings."

BTU Buddy says, "Let's put some gauges on the pump and see what it says."

Bob puts some water gauges on the inlet and the outlet to the pump and says, "The inlet is in a vacuum and the outlet is reading 50 psig. You were right, the vacuum would pull in air if the seal leaks."

BTU Buddy says, "Notice that there's a valve at the pump inlet and the outlet. The inlet valve is partially closed. Someone has tried to limit the flow across the pump. Let's get an ammeter and see what the amperage is."

Bob clamps his ammeter around one of the pump wires and it reads 45 amps. The full load amps for the motor is 45 amps, so the motor is running at full load.

BTU Buddy says, "Watch the ammeter when you open the suction valve to the pump."

Bob begins to open the valve and the amperage begins to rise above full load.

"Now close the outlet valve and bring the amperage down," BTU Buddy says.

Bob asks, "Won't the amperage rise if I start to close the pump discharge?"

BTU Buddy explains, "It doesn't matter how you reduce the flow across a centrifugal pump, you can either use the inlet or the discharge. When you reduce the water flow, the amperage will reduce," Figure 50-4. "It would be different if it were a positive displacement pump; you wouldn't be able to shut down the outlet without causing the amperage to rise."

Bob asks, "How would I know the difference?"

BTU Buddy says, "Look at how the water enters the side of the pump. If you look at the pump housing you can see that it has an impeller inside."

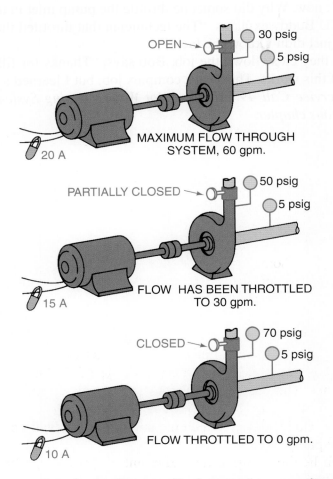

**FIGURE 50-4.** This illustration shows that throttling a centrifugal pump at the pump outlet causes the amperage to go down. You would think that closing off the outlet would cause an amperage rise, but it is actually water flow through the pump that causes the amperage. More flow causes more amperage.

Bob begins to close down the outlet valve and the amperage drops back to full-load amps.

"Now see if you can control the pump to full-load amps by opening the inlet valve all the way and closing down on the outlet valve," BTU Buddy says.

Bob keeps experimenting until the inlet is wide open and the outlet is partially closed.

BTU Buddy says, "You now have the same water flow. What's the inlet pressure?"

Bob looks and says, "We now have 10 psig of inlet pressure and 55 psig of outlet pressure. If the pump running in a vacuum was causing a leak, it won't now. Why did someone throttle the pump inlet in the first place?"

BTU Buddy explains, "The technician that throttled the pump inlet probably just didn't know the facts."

As they close down the job, Bob says, "Thanks for filling me in on how all of this works. This was a complex job, but I learned a lot."

*"Service Call 49: Air in a Hot Water Heating System" appears in the previous chapter.*

# Stolen Outdoor Heat Pump Unit

Bob receives a call from the dispatcher saying that a small retail store that uses a heat pump is without heat. The weather is cold and Bob has to finish another job, so he calls the store owner and tells him to switch to emergency heat until he can get there.

Bob finishes up at the first call and drives over to the store with no heat. He goes in and finds the owner in the stock room, where the heat pump indoor coil is also located. Bob asks the owner, "What seems to be the problem?"

The owner responds, "I arrived early this morning to get things in order to open and noticed that it was cool in the store, so I checked the thermostat. I found that it was set at 72°F, but the store temperature was 60°F. I called your company and they said they would contact you. When you called, I did what you said; I set the thermostat to emergency heat. We began to feel the heat in a few minutes, but it's still chilly in the store. Would you mind explaining 'emergency heat' to me?"

Bob explains, "A heat pump is made up of two heating systems. The actual heat pump uses the outdoor unit to absorb heat from the outside air, and is very efficient," Figure 51-1. "For every watt of electricity it consumes, it pumps up to 3 watts of heat into the store. It's able to accomplish this by absorbing heat from the outside air. When it begins to get very cold, for example below about 35°F, the heat pump's efficiency begins to diminish and it needs auxiliary heat. The auxiliary heat for this system is an electric furnace that assists the heat pump in cold weather. When the electric furnace has input of 1 watt, it puts out 1 watt of heat. It only runs when the heat pump can't hold the store temperature at the set point. From that explanation, you can see that you don't want to run the electric heat any longer than you have to.

"The electric heat has a feature called 'emergency heat' that will shut the heat pump off and run on just electric heat in case of an emergency, such as you had here this morning," Figure 51-2.

Bob says that he'll get started right away and get to the bottom of the problem.

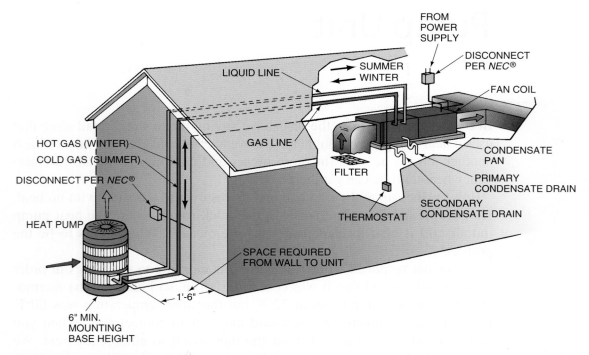

**FIGURE 51-1.** The outdoor unit absorbs heat from the outdoor air. Notice that the only thing connecting it to the structure is piping and wiring.

**FIGURE 51-2.** This thermostat has both an emergency heat and auxiliary heat feature that shows the customer what is operating.

He goes to the air handler and checks the filters; they're clean. Next he goes to the room thermostat and sets it to regular heat so the heat pump will run.

He goes to the door leading from the storeroom to the back of the store, where the heat pump outdoor unit is located. When he opens the door, he can't believe what he sees. There is no outdoor unit! He calls to the store owner to take a look. They go outside, and where the unit used to be located, there's only the base, but no unit. The store owner says, "It looks like someone took an ax and just cut the pipes off."

Bob says, "They turned off the power and cut it loose. I guess they just set it in the back of a pickup truck and off they went."

The owner says, "I've read that copper is priced so high now that people are stealing units just for the copper."

Bob goes back inside and turns the thermostat back to emergency heat. Split system heat pumps have two power supplies, one for the outdoor unit and one for the indoor unit, which contains the control circuit. That's why the system could still operate without the outdoor unit. This is going to take a while.

About this time, BTU Buddy appears and asks Bob, "Now what?"

Bob says, "I'm not sure where to start. I know that we need a new outdoor unit, and that I'll need to match it to the indoor unit. I guess I should take down the model number of the indoor unit."

BTU Buddy says, "This unit looks to be at least 10 years old, if not more. It's probably less than a 10 SEER unit. You'll have to replace the indoor unit as well as the outdoor unit to bring it up to the new 13 SEER standard. You'd better explain this to the owner, as insurance will pay for at least part of it, but he'll have to pay for part of it, too. I'd suggest that the owner call his insurance adjuster and get some advice."

Bob goes to the owner and explains the situation to him. The owner calls the police and the insurance adjuster to get things started.

As Bob looks over the job, BTU Buddy asks, "How do you think you can prevent this from happening again?"

Bob says, "I hadn't even thought about that. I wonder if a chain-link fence would help?"

BTU Buddy says, "That just makes them have to go through one more step; breaking the lock. You may want to consider moving the unit to the roof of the store. It would make more noise inside, but it would be over the storeroom. The electrical and piping would have to be rerouted to the roof. When you give the owner a proposal, you should show it both ways."

Bob gathers the information, goes to the office, and lets the sales force work up proposals for an installation on the roof and behind a fence. He goes back to the store and explains it to the owner.

The owner decides that he likes the idea of a fence better because it doesn't involve the roof. He explains, "I've had roof problems before, so I would rather stay away from there. Service would also be easier behind a fence. I'll explore getting an alarm system that connects to the unit so that when it's serviced, you'll have to shut off the alarm. The lock and an alarm may be enough to stop another incident. It's a shame that someone will only make a few dollars on the copper sold, yet it will cost several thousand to replace the system."

Bob says to the owner, "I'll get the wheels turning on this and get it started."

As they drive away, BTU Buddy says, "Bob, you really handled that like a professional. Good job."

Bob replies, "I really like the idea of developing good customer relationships. You have proven to me that getting and keeping a loyal customer is one of the greatest secrets to being a professional service technician. Thanks for your help."

# Grounded Compressor Motor

The dispatcher calls Bob and says, "A customer called and his heat pump is not running. He reset the breaker at the outdoor unit several times, and it just tripped again."

Bob tells the dispatcher to call the customer back and tell him to switch the unit to emergency heat and to not reset the breaker again.

When Bob arrives, the owner tells him that the unit was working just fine last night but not this morning when he got up. Bob explains that he should have only reset the breaker one time, as resetting can stress it.

Bob then gets to work. He goes to the outdoor unit and finds the breaker tripped, so he turns it to the off position. Since it has been reset several times, he decides to check the voltage at the load side of the breaker. He checks it line-to-line and line-to-ground. It shows 115 V on one line-to-ground. It's a good thing that he checked it; he could have received an electrical shock.

Bob goes to the owner and explains that he has to turn off the main breaker for a few minutes because the breaker on the unit is defective. He wanted to warn him that this would shut down his computer if it were operating. The owner asks to have 5 minutes to shut down the computer. When he returns to tell Bob that the computer is off, he asks, "Is that the only problem?"

Bob says, "I don't know. I have to be able to work without any power to the unit to determine that."

Bob goes to the main panel and turns off the main breaker. He changes the defective breaker at the unit and turns the main breaker back on and notifies the owner that he's through with the main circuit.

He's now ready to troubleshoot the unit. With the breaker to the unit off, he places his ohmmeter leads on the load side of the breaker and sets the ohmmeter to RX1. The reading is infinity. All looks good so far.

He removes the panel to the unit's electrical components and pushes the armature to the contactor to close the contacts, Figure 52-1. The meter reading is 10 ohms. He places a meter lead on the load side of the contactor, and the other to a ground source. The meter reads 0 ohms, a dead short to ground. Now he's getting somewhere, but he needs to figure out where the short is located.

He removes the compressor terminal cover and removes the wires. He places the leads from common to run, and the meter reads 10 ohms. Then he places a lead on the compressor run terminal to ground, and the meter shows 0 ohms resistance. The compressor was shorted to ground, Figure 52-2. It could just as easily have been the fan or the wiring, but this narrows it down to the compressor.

Now that Bob has found the problem, he tells the owner that he'll pick up the necessary supplies and be there in the morning. As he's picking up his tools, BTU Buddy arrives and says, "How are you going to leave the unit until tomorrow?"

Bob says, "I'll just put the panel back on and leave the breaker off."

BTU Buddy says, "If you leave the power off to the crankcase heater, the compressor will be full of refrigerant. The refrigerant will mix with the oil, and recovery will be hard to accomplish. Whatever you do, leave the crankcase heat on. Tape the compressor leads where they'll be safe and turn the breaker back on."

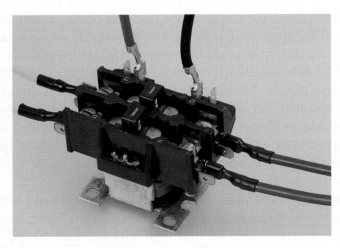

**FIGURE 52-1.** This contactor has the line voltage coming in one side and the load voltage to the compressor out the other side. The two wires standing straight up are the 24 V control wires.

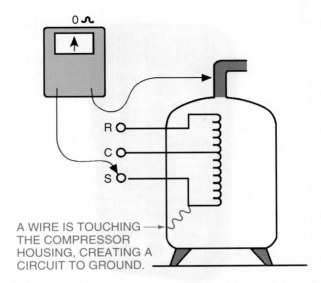

**FIGURE 52-2.** This grounded compressor only shows 0 ohms to ground.

He adds, "What supplies do you need for tomorrow?"

"A compressor and refrigerant," Bob replies.

"What about the contactor and the run capacitor?" BTU Buddy asks. "They were stressed also."

Bob says, "I'm not really thinking this all the way through, am I?"

BTU Buddy says, "No, there's a lot to be considered when a motor is shorted. For example, you don't know if it's a motor burn, or just a wire shorted to ground."

"What should I do next?" Bob asks.

BTU Buddy says, "You could run an acid test on the system to determine if there are acids present. Do you have one of those acid test kits that you can pass some refrigerant through for a test?"

"No," Bob says.

"Okay," BTU Buddy says. "We'll just have to wait until we take the old compressor out and check the piping for signs of burned oil and smell. We may need a suction-line acid removing filter in addition to a liquid line filter-drier. Put them on the list; we'll use both if needed."

Bob checks the capacitor with his capacitor tester, Figure 52-3. It's in good shape. Then he visually checks the contacts, which look bad, Figure 52-4, so he adds a contactor to the list of supplies.

**FIGURE 52-3.** A capacitor tester can be used in the field to determine the capacitance of a capacitor. *(Courtesy Davis Instruments.)*

**FIGURE 52-4.** Good contacts are shown next to contacts that are pitted and should be replaced. *(Courtesy Square D Company.)*

BTU Buddy then says, "I think you have all the bases covered. Leave the power on and replace all of the covers."

As they drive away, Bob says, "There are so many details to doing a job correctly. How do you ever remember them all?"

BTU Buddy says, "Well, you remembered one of the first details. You warned the owner that you were going to shut off the power and that he should shut down the computer. Some technicians would just shut off the power, and if the owner were uploading or downloading business files and had no battery backup for the computer, it could cause problems for him. These things become instinct as you gather more experience. You're doing well with your training."

Bob says, "Thanks, but I feel as though I still need a lot of training."

BTU Buddy says, "Training is a forever thing in a good technician's life."

*Look for Part 2, "Service Call 53: Grounded Compressor Motor (continued)," in the next chapter.*

# Grounded Compressor Motor (continued)

Bob and BTU Buddy have gotten together to finish a job from the previous day—a grounded compressor motor that needs to be replaced. Bob uses his recovery machine to recover the refrigerant from the system, but when he connects the system to the gauges, he notices that the refrigerant in the system smells really bad. BTU Buddy says, "That's a sign of bad motor burn. I'm glad you have a suction-line acid removing filter."

He goes on, "The system is in a vacuum after the recovery. What are you going to do next?"

Bob says, "Just open the valves and let air in. We're going to evacuate the system anyway."

BTU Buddy says, "Turn off the power first and lock the disconnect switch. Then the system should be brought back up to atmospheric pressure with dry nitrogen. If just air is let into the system, the oxygen from the air will mix with any acid in the system and could cause corrosion, so it's a good practice to fill the system with nitrogen," Figure 53-1. "Recovering the refrigerant and pulling the system into a vacuum removes all refrigerant, but acid will remain in any oil that's left in the system coils and piping."

Bob goes to the truck for his nitrogen cylinder and regulator. He lets nitrogen into until the system is back to about 5 psig of pressure. Then he removes the gauges and lets the system equalize with the atmosphere. "It's time to cut the compressor loose and remove it," he says.

Bob then loosens the hold-down bolts to the compressor base and removes them so it can be lifted from the cabinet. He had already disconnected the wiring.

Now it's time to cut the compressor loose. BTU Buddy says, "Look at this diagram before you open the system, and let's plan all of the moves ahead of time," Figure 53-2. Notice that the only permanent suction line is between

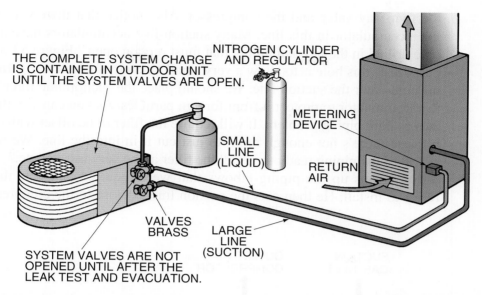

**FIGURE 53-1.** This illustration shows the gauges connected to the refrigerant cylinder with the nitrogen cylinder ready to connect.

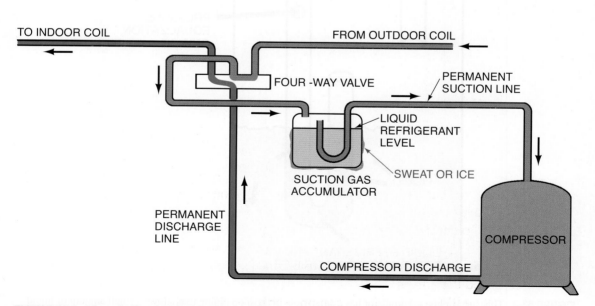

**FIGURE 53-2.** The only line that remains a permanent suction line is the line between the four-way valve and the compressor. The filter must be located in this line. Any other location would subject the filter to high pressures during one of the heat pump cycles.

the four-way valve and the compressor. Also notice that there's a suction-line accumulator in this line. Many suction-line accumulators have a small hole drilled in the internal piping that must remain open," Figure 53-3. "The purpose of this hole is to allow a very small amount of liquid refrigerant and oil to re-enter the suction line. We should place the suction line filter before the accumulator to protect it from foreign particles. As you can see, this creates a slight piping problem. It will require the filter to be offset with elbows because there's not enough room to just cut it in into the line. We want to get all of the piping ready to install the filter at the very last minute."

Bob cleans up the piping elbows that were required and gets the pipe ready to install. He then cuts the suction line as close to the compressor as

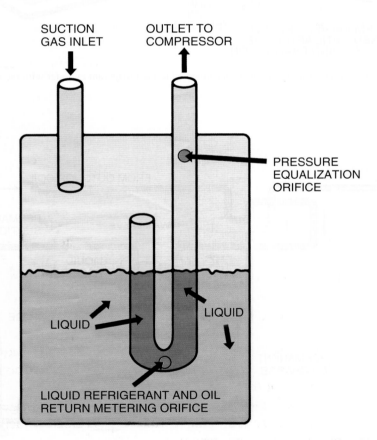

**FIGURE 53-3.** This suction line accumulator has a small hole in the loop piping that allows a small amount of liquid refrigerant and oil to return to the compressor. This is the only way oil can return when it gets into the accumulator, and it helps move any liquid refrigerant that may be in the accumulator to the compressor without risking liquid slugging.

he can, and cuts the discharge line close to the compressor. He then sets the burned compressor out.

BTU Buddy says, "Examine the suction and discharge lines internally and tell me what you see."

Bob looks them over; he even wipes the inside of the lines with a clean white rag. He says, "There's black oil in the discharge line, but the suction line is clean."

BTU Buddy says, "It looks like the compressor burned while running, so the contamination must have moved out the discharge line. The liquid line drier will get most of it and the suction line filter will protect the accumulator and compressor."

Bob sets the new compressor in place and starts cutting the pipe to the correct length to install the filter. As he reaches for the filter, BTU Buddy says, "Let's get the piping prepared to install the filter, and I'll show you something just before we open the filter to the atmosphere. Now cut the liquid line and get it ready for the filter-drier, but don't open it yet."

Bob gets everything ready for the filters and asks BTU Buddy what is next.

"Install your gauges on the system gauge ports and connect the nitrogen cylinder."

Bob does this and turns to BTU Buddy, who says, "Now purge nitrogen through the entire system at 100 psig."

Bob turns the nitrogen up to 100 psig and lets it purge though the system for about 10 seconds.

BTU Buddy then says, "All vapors in the system are as clean as we can get them without running the system. Now is the time to install the filters while letting the nitrogen trickle through the system. Notice that the liquid-line filter is a two-way filter. The liquid changes direction from heating to cooling, so the two-way filter-drier is necessary," Figure 53-4.

Bob sets the nitrogen to barely flow into the system while he places the liquid line drier in the line. He then connects the piping to the suction line filter.

BTU Buddy says, "Disconnect the nitrogen cylinder and open the system to the atmosphere, and use the torch to fasten all piping connections."

Bob does all of the torch work and BTU Buddy says, "Now turn off the gauge manifold valves as quickly as you can. You've heated the piping and as it cools, the nitrogen inside it will cool and create a vacuum within, pulling atmosphere into the system.

"You have put this system back together with nitrogen inside, which will keep down oxidation at the points where you used the torch. You have

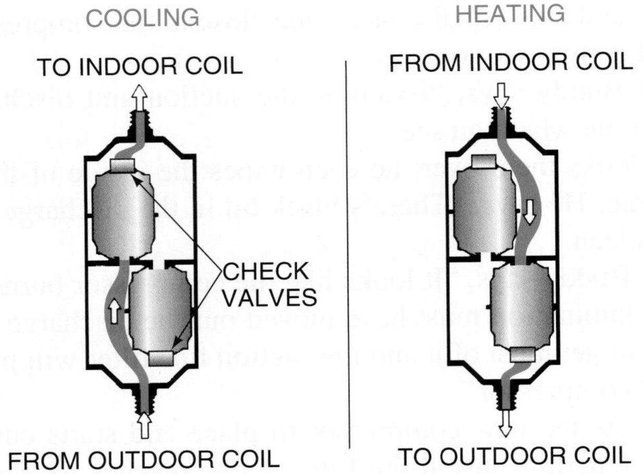

COOLING                    HEATING

TO INDOOR COIL          FROM INDOOR COIL

CHECK
VALVES

FROM OUTDOOR COIL       TO OUTDOOR COIL

**FIGURE 53-4.** This liquid line filter-drier allows it to filter in either direction by means of check valves. If just a plain filter is used, it will filter in one direction and particles will be cleaned from the filter and circulate again when the direction changes.

placed the filters and drier into the lines that weren't contaminated with air and oxygen by being open to the atmosphere for more than a moment, so they'll have the most capacity to do their work efficiently. It's time to put R-22 into the system up to about 10 psig, and then use the nitrogen to pressure the entire system up to 150 psig and leak-check all connections, including your gauge connections."

Bob asks, "Why don't we pressure the system up to 250 psig?"

BTU Buddy says, "You should only pressure the system up to the working pressure of the weakest link in the system. That would be the compressor housing. Remember, the compressor can pump several hundred pounds of pressure, but the compressor shell is a low-side device. The discharge line that handles the high pressure goes from the compressor inside the shell, through the discharge line to the outside of the shell in this reciprocating compressor. Most compressor shells have a 150 psig working pressure. It is best not to exceed that pressure for the compressor shell."

Bob leak-checks the system, evacuates it to a deep vacuum, and charges it by measuring the charge into the system. Then he starts the system.

BTU Buddy says, "It's nice and cold, so the system should get some long running hours in, which is good. The refrigerant in the system is a great solvent. It will clean any burnout residue from the coils and piping, and be deposited in the acid-removing filter and filter-drier. There is plenty of filter capacity here so there's nothing to worry about. This system should give plenty of good service for many years. Oftentimes a technician will take shortcuts, and some acid gets left in the system that will eventually cause it to fail again. This system was professionally cleaned to the best standards."

Bob says, "There's a lot to this work. It's very technical but every step has a logical reason. Thanks for giving me the best."

*Part 1, "Service Call 52: Grounded Compressor Motor," appears in the previous chapter.*

# Heat Pump with a Stuck Check Valve

Bob receives a call from the dispatcher that a regular customer seems to be having a problem with his heat pump. The weather is cold and the unit's auxiliary heat light is staying on most of the time.

Bob arrives at the job and talks to the customer, who is an engineer and really keeps a close eye on how his heat pump operates. He tells Bob, "The auxiliary light at the thermostat almost never goes off, and when it does go off, the heat pump continues to run, but the air isn't warm like it used to be."

Bob goes out to the heat pump and does a touch test on the lines. The hot gas line is cool. He removes the compressor compartment door and feels the compressor. It's cold all over, even the crankcase, which should be warm no matter how cold it is outside. From the touch test, Bob believes the compressor has liquid refrigerant entering the suction line. He observes that it's a reciprocating compressor and there's an accumulator in the permanent suction line (between the four-way valve and the compressor). He calls the office to see if anyone else has worked on this unit before. The dispatcher pulls up the records and tells Bob that he is the only one who has serviced the unit.

Bob is sure that the unit has too much refrigerant and he can't figure out why. He goes to the house and talks to the owner, who assures Bob that no one else has serviced the unit. Bob tells the owner about the symptoms before going back to the unit and applying gauges.

The gauge reading shows a high suction pressure and a low discharge pressure. This looks like a defective compressor now. Bob's head is really spinning.

BTU Buddy shows up about this time and asks Bob what he thinks.

Bob says, "The compressor is getting liquid from somewhere but from what the records indicate, I don't think there's too much refrigerant in the

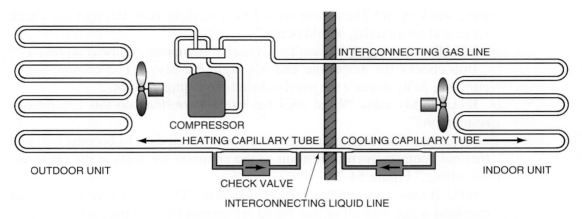

**FIGURE 54-1.** A diagram of a heat pump cycle with two capillary metering devices with check valves to make the refrigerant flow through the correct device.

system—it has always performed correctly. The symptoms indicate a low charge, but it turns out to be an internal problem."

BTU Buddy says, "Explain the refrigerant circuit for this unit in the heating cycle."

Bob says, "Let's start with the compressor discharge." He produces a diagram from his truck, Figure 54-1. "The hot gas leaves the compressor through the hot gas line at the bottom of the compressor, then proceeds through the four-way valve to the indoor coil, where hot gas is condensed to a liquid and leaves at the bottom of the coil. The liquid line carries the liquid through the indoor coil metering device and its check valve, which is open with this direction of flow. The liquid then flows to a metering device at the outdoor coil where there is a check valve in one line and a metering device in the other line. The liquid cannot flow through the check valve and so is forced through the metering device, and is then metered into the outdoor coil where it absorbs heat from the outdoor air. The liquid refrigerant in the outdoor coil is boiled to a vapor, and leaves the outdoor coil to return to the compressor as vapor. The compressor then compresses the vapor to a hot gas, to go around the circuit again."

## THE CRITICAL QUESTION

BTU Buddy then asks the critical question, "Bob, explain again what happens at the outdoor metering device."

Bob says, "The liquid refrigerant is forced through the metering device by the check valve. I see what you're getting at—what if the check valve were stuck open? Then there would be a straight flow through the check valve and no metering would occur."

BTU Buddy says, "Why don't you check the amperage at the compressor?"

Bob checks the amperage and it's high. He says, "The compressor is pumping. Why doesn't the overload shut the compressor off?"

BTU Buddy asks, "What does the internal overload on this compressor respond to?"

Bob says, "The high amperage of the motor causes it to become hot, and thermal contacts open the circuit to the common terminal in the compressor wiring," Figure 54-2.

BTU Buddy asks another critical question, "Do you believe the internal overload is hot with all of that liquid refrigerant flowing over it?"

Bob says, "I see what you mean. It's running overloaded but cool. Is there anyway to repair the valve?"

BTU Buddy says, "Sometimes you can tap on it with a soft face hammer and reverse it a few times while tapping on it. That may free it up, but this will likely occur again. The valve is either dragging internally from wear or it has a particle stuck in it. The best option is to replace it and forget it."

Bob goes to the owner and explains the situation. The refrigerant will have to be removed from the unit to make the repair. The owner says to go ahead with what needs to be done.

BTU Buddy makes a suggestion. "Turn the unit to emergency heat, which will shut off the compressor. Let the unit sit like that, with the crankcase heat on the compressor overnight, to boil the liquid refrigerant

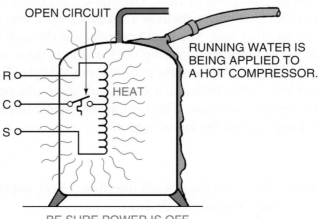

FIGURE 54-2. This illustration shows a compressor that is hot with the internal overload in the open position.

out of the compressor crankcase. Then come back tomorrow and remove the refrigerant and change the valve."

Bob then tells the owner the plan of action.

Bob returns the next morning and recovers the refrigerant. He then changes the valve and adds a liquid line filter-drier. He evacuates the system and weighs in a measured charge that includes the extra amount for the new filter-drier. He then starts up the heat pump. The hot gas line gets hot and Bob goes into the house to check the air temperature with the auxiliary heat turned off. The air temperature is 100°F, so Bob explains to the owner that the system is operating correctly.

"Thanks for a professional job well done," the owner says.

Bob is driving away when BTU Buddy says, "Good job, Bob. The owner was impressed with the way you did your work."

# An Oil Furnace with Sooted Electrodes

The dispatcher calls Bob on a cold morning and explains that a customer's oil furnace will not fire on a regular basis. The customer says that sometimes the furnace starts, and sometimes it doesn't, in which case he has to reset the primary control. Bob calls the home, and when the wife answers, he tells her not to reset the furnace until he gets there, within the hour.

When Bob arrives, the wife tells him that she has not reset the control again. He asks her when the burner was last serviced and she says about 2 years ago.

Bob then says, "I know it sounds like a sales pitch, but oil burners benefit from service every year. They're not like other heating equipment, and really need to be checked regularly."

Bob goes to the garage where the furnace is located and turns the furnace switch off. He uses a flashlight and mirror to inspect the combustion chamber for excess oil. There is no oil in the chamber, but what he sees is a very dirty combustion chamber. It has a lot of soot in it from unburned oil.

He decides to start the furnace and see what it does. He goes to the thermostat and sets it to call for heat, but he needs to turn the furnace switch back on before it will start up.

He goes to the furnace, holds the inspection door open, and turns on the switch. The furnace fan and pump motor start, but it doesn't ignite for a few seconds. When it does ignite, a puff of black smoke comes out the inspection door and it makes a slight booming sound. The burner continues to burn, but the fire is very orange and there's a lot of smoke. He steps outside to look at the flue and there's smoke in the flue gas.

Bob turns the furnace off and is gathering his thoughts when BTU Buddy shows up and asks, "What's going on, Bob?"

Bob explains what the furnace is doing, but that he doesn't see any reason for the furnace not to fire each time.

BTU Buddy says, "Why don't you remove the burner and give it a good examination. You'll need to remove it to service it anyway. Did you ask the owner how long it has been since it was serviced?"

Bob responds, "She said it's been about 2 years."

BTU Buddy says, "The timing is right to do a full service on the system."

Bob brings his tools in from the truck and removes the burner section of the furnace. He looks it over closely.

BTU Buddy asks, "What do you think?"

Bob says, "It looks like the fire has been burning with some smoke for a long time. The electrodes should be white, but they're black," Figure 55-1. "I'm going to start by cleaning them up and changing the oil nozzle."

"That sounds good, you're on the right track," says BTU Buddy.

Bob takes the electrodes off and starts soaking them in solvent to get them clean. He uses the mirror to inspect the combustion chamber for damage. There is some soot, but the furnace combustion chamber and heat exchanger look good.

Bob cleans the electrodes and they become almost white again. BTU Buddy says, "I think the electrodes were so dirty that they were leaking an arc-to-ground part of the time, and the furnace wasn't firing as it should. Electricity will follow the path of least resistance and the direction of the arc may have depended on whether the electrodes were hot or cold. If it doesn't arc across the electrodes, the burner won't fire."

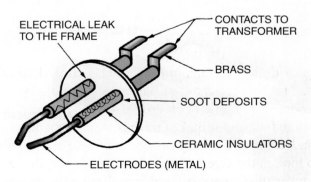

**FIGURE 55-1.** When the electrodes have soot on them the high voltage electrical arc may follow the soot to ground. Soot is made of carbon and carbon is a good conductor of electricity.

Bob changes the oil filter and fills the filter cartridge with fuel oil to prevent air in the lines. Then he changes the nozzle and keeps the nozzle line from getting air in it.

After placing the electrodes back into position, he uses an adjustment tool to set the electrode distance, Figure 55-2. Then he puts the burner back together and places gauges on the pump to see the oil pressures. The tank is below the burner, so it's a two-pipe system. He's ready to start the system.

He uses the furnace switch to start the system, and it fires up correctly. The gauges read 2 inches of Hg (mercury) inlet and 100 psig outlet. The pressures are good.

Bob says, "The fire looks good now. I guess the furnace burner just needed servicing." He starts to pick up his tools when BTU Buddy says, "Don't you think you should finish this professional service call by doing a flue gas analysis?"

Bob says, "Okay. You aren't going to let me get by with anything are you?"

BTU Buddy notes, "You're being paid to do a complete job, so let's do it."

Bob gets out his combustion analyzer, Figure 55-3, and after the furnace is up to temperature, he runs a test. The burner is well within the proper range, so he's finally through.

Bob explains to the owner what he has found and what he has done. He also reiterates that the furnace should be examined each year at the beginning of the season to make sure it's performing correctly.

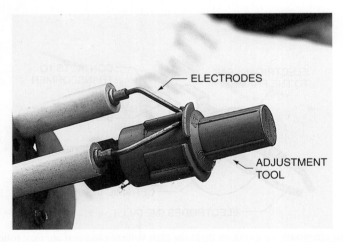

**FIGURE 55-2.** This electrode gauge is designed for setting the electrode tolerances.

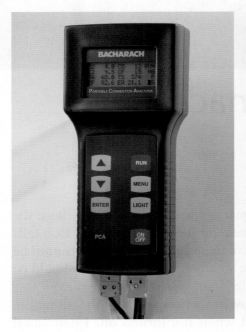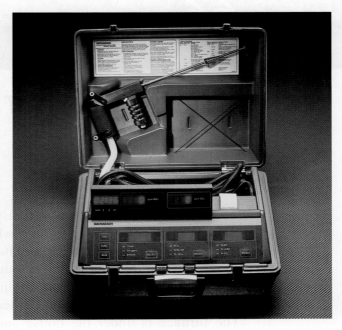

**FIGURE 55-3.** Electronic combustion analyzers from Bacharach, Inc. *(Courtesy Bacharach, Inc., Pittsburgh, PA USA)*

She agrees and thanks him for doing a good job. Bob then removes the gauges and puts his tools away.

As they're riding away, BTU Buddy says, "I'm glad you ran that combustion analysis. Without if you wouldn't have known that the furnace was performing correctly. Now you have it written in your report that you did the complete job."

# Smelly Gas Furnace

Bob has received a call from the dispatcher to go to a house where the residents smell some kind of fumes when the furnace is running. Bob tells the dispatcher to call them back and have them turn the furnace off until he can get there, within the hour. Since the weather isn't too cold, it should be okay for the furnace to be off for a while.

Bob arrives and talks to the homeowner who describes the situation, "The furnace is under the house and when it's turned on, I can smell fumes that don't seem normal. The furnace is old, but it's supposed to be very high quality."

Bob gets his flashlight and goes under the house. The furnace is a horizontal model, mounted on a sturdy block foundation. The vent pipe runs about 8 feet to a chimney that goes to the top of the two-story house. The furnace looks old, but is in very good shape. Bob examines the flue connector that runs from the furnace to the chimney and discovers the problem—it's rusted so badly that it's falling apart. He measures from the furnace to the chimney.

Bob removes the doors to the furnace and uses a mirror to examine the heat exchanger, which looks good. He removes the draft diverter where the flue pipe is connected, Figure 56-1, so that he can see down into the heat exchanger. Everything looks fine.

Bob explains the problem to the homeowner and starts to leave for the shop to pick up materials to replace the vent pipe, when BTU Buddy appears and asks, "What are you going to do, Bob?"

Bob says, "Oh no, I must be about to do something wrong. That's when you always seem to appear. I'm going to the shop for a vent pipe for this furnace."

BTU Buddy asks, "What kind of pipe are you going to get?"

"One just like the one that's on the furnace now—it's lasted for many years," Bob says.

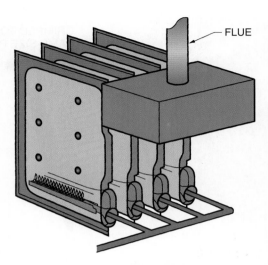

FLUE

**FIGURE 56-1.** The flue pipe is attached to the box called the draft diverter. When the diverter is removed, you can see down into the heat exchanger.

Then BTU Buddy says, "The one that's on the furnace is a single-wall vent pipe, and that's the reason it has rusted out. It really should be replaced with a double-wall vent," Figure 56-2.

Bob says, "I was just going to replace it with what had been used before."

BTU Buddy says, "The current code calls for double-wall pipe because single-wall pipe will fail in this application. It will take a long time to fail, but it will fail."

"What's the big difference?" Bob asks.

BTU Buddy explains, "The flue gasses leaving the furnace contain a lot of moisture. If the flue gasses cool down below the dew point temperature of the flue, some moisture will condense in the flue pipe. The temperature under the house is cool, probably 50 or 60°F—well below the dew point temperature of the flue gasses. The moisture is slightly acid, about like a soft drink, and it will corrode the pipe to the point that it will fail, just like the flue at this job. Condensate is what caused it to fail. Also, the flue gasses will be cool and slow to start rising up the chimney. You want the flue gasses to rise as quickly as practical to prevent flue gas from spilling out under the house. Double-wall pipe will keep the flue gasses warm enough to get them to the chimney and they will rise up and out to the atmosphere."

Bob then says, "I've worked with double-wall pipe in the past, and it can be difficult. Everything needs to be just right."

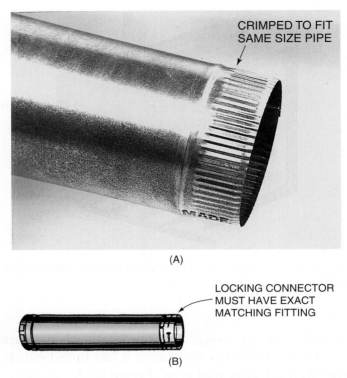

**FIGURE 56-2.** (A) Single-wall vent pipe. (B) Double-wall vent pipe.

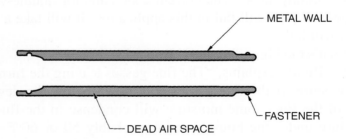

**FIGURE 56-3.** This illustrates the air space between the inner wall and the outer wall of double-wall vent pipe.

BTU Buddy goes on to explain, "Double-wall pipe has a dead air space between the inner and the outer pipe," Figure 56-3. "This dead air space has an insulating effect. Joining it using the special connections that lock together is part of keeping the integrity of the vent system. You'll notice that you don't need sheet metal screws to fasten the pipe connections together, as they have their own locking system. In fact, you should never use screws to connect the pipe, except where the pipe fastens to the furnace collar," Figure 56-3.

"So, what should I do?" Bob asks.

BTU Buddy says, "Tell the customer what is required to get the job up to the code level by telling him about the double-wall pipe."

Bob goes back to the customer and explains what has to be done. He says, "The fumes you smelled are called aldehydes, and are present in the flue gas. They aren't poison, like carbon monoxide, but they do alert you to flue gas being present. It's much like the burners on your gas stove; you smell the aldehydes when you light the stove, but you have a nice blue flame and it's safe. The aldehyde odor is very slight, but noticeable to someone with a keen nose. A yellow flame will contain carbon monoxide and smell the same, like the aldehydes."

Bob explains what needs to be done and the customer says, "By all means, get the system up to code."

Afterwards Bob installs the pipe, starts the furnace and checks the flame, oils the motor, and replaces the air filter.

BTU Buddy says, "There's one other aspect to this job that you didn't notice, but we can talk about it over lunch tomorrow. This job is now up to code and all is well."

As they drive away, BTU Buddy comments, "The customers all seem to want to do the safe thing—you did the right thing by bringing it to their attention."

"Would most service technicians go to double-wall pipe in a situation like this?" Bob asks.

BTU Buddy says, "Some would just replace the vent with single-wall pipe and it would work well for a while, but we can't be sure how long. It pays to be safe."

# Luncheon about Flue Vents

BTU Buddy and Bob have met for lunch and Bob has a question about the last service call yesterday. "You said that we had some unfinished business about the service call yesterday, but you also said the call was complete. What did you mean?"

BTU Buddy says, "You did a good job replacing the single-wall flue pipe with double-wall pipe on that horizontal gas furnace. Any time a vent passes through an un-conditioned space, double-wall vent must be used for proper venting. You got the system up to code level, but what you didn't do is examine the chimney to see if it was safe to vent into."

Bob asks, "What do you mean by that? It's a chimney."

BTU Buddy explains, "I looked at the chimney and saw that it was safe for venting; what I wanted was for *you* to examine it and tell me that it was safe."

"What's the difference?" Bob asks.

BTU Buddy explains, "Some older chimneys are just masonry, brick, or concrete blocks with no lining. They tend to deteriorate over time and actually can fall over."

"How in the world can that happen?" asks Bob.

"These older chimneys were likely used for coal and wood, and have a coating of soot from back then. They're very slow to warm up and a lot of condensate can form before they get hot. This condensate, as I mentioned before, is slightly acid and will work with the old soot to actually eat the mortar out from between the brick," explains BTU Buddy, Figure 57-1. "The chimneys being built today have ceramic liners with a glaze, much like the glaze on a coffee cup. If there is condensate, the glaze can stand up to the mild acid. The liner is not directly against the brick, so there's a dead air space between the brick or block and the liner, that acts as an insulator. Only the liner has to get hot for the flue gasses to rise. When the chimney is lined, you can back away from the house and see the liner protruding above. The liner is light brown or red," Figure 57-2.

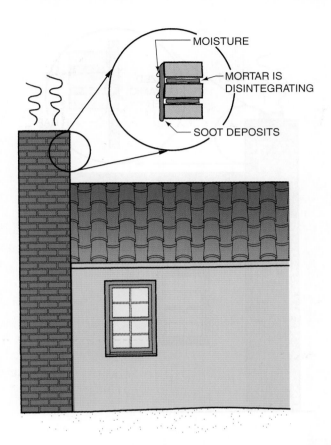

**FIGURE 57-1.** The combustion gas has a lot of moisture in it, which is slightly acid. When combined with the soot in an old chimney there is enough acid to attack the mortar between the bricks and break it down.

Bob asks, "What would happen if the chimney were the only way to vent a furnace but it was the type that you aren't supposed to use for a vent? Would you just use it anyway?"

BTU Buddy explains, "There's often an alternative. Some chimneys can be lined with a stainless steel liner. It's worked down the chimney and connected to the flue from the furnace," Figure 57-3. "This is not a simple installation; it's one for the experts who do this kind of work."

"There sure is a lot to venting systems. It seems complicated," Bob says.

BTU Buddy says, "The National Fuel Gas Codes are used for venting practices and are available in the library, or your company probably has the books in the company library. You should check them out and read through them. There are a lot of illustrations to guide you. When you approach any job, you need to do an inspection to see that it meets the code specifications."

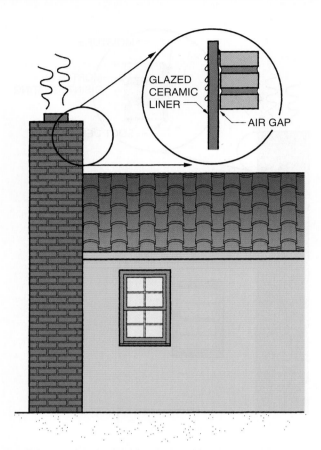

**FIGURE 57-2.** This chimney has a ceramic liner in it that will not deteriorate with moisture. There is an air space between the liner and the brick that acts as an insulator so the flue gets up to temperature faster.

Bob asks, "Is there anything else that I should know about the job?"

BTU Buddy says, "Yes. There is a relationship between the length of the horizontal run and the vertical rise for vents. I think it's pretty easy to see that if you had a long lateral (horizontal) run and a short vertical rise, the flue gasses would not vent as well as if you had a short lateral (horizontal) run and a long vertical rise. The house on the service call yesterday was two stories high and had a lateral run of 8 feet. That looked good to me. There are tables that show this relationship," Figure 57-4. "Roughly, the lateral run should be no more than about 25% of the total run. The house yesterday had 8 feet of lateral run, 2 feet of chimney under the floor, 2 floors with 9-foot ceilings, and an attic of about 8 feet. Add to this the rise above the roof of about 8 feet, and you have 44 feet, which would allow a lateral run of about 10 feet. To be sure, you should check that against the code book."

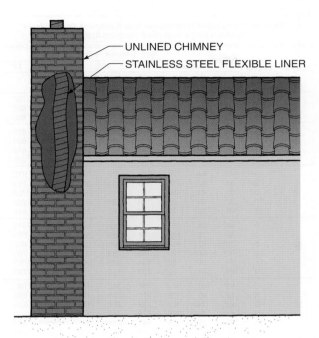

**FIGURE 57-3.** This old chimney has no lining and the furnace is vented through a stainless steel liner that is worked down through the old chimney and connected to the furnace.

Bob says, "There's so much to remember."

BTU Buddy says, "That's for sure, and the longer you're in the field, the more information you gather. It's important to always keep the safety of the customer in the forefront of your thinking. Protect the customer, and it's hard to go wrong. You should always look at the safety aspects of any job and record any problems you see, and then let the customer sign the ticket. This could eliminate problems for you and your company if anything ever happens and you've documented it and made the customer aware of it.

"You should also be aware of how the vent terminates above the house. The code says that the terminal point of the vent should terminate 2 feet above any portion of the roof that is within 10 feet of the flue. This is to keep winds from coming over the structure and creating a positive pressure above the roof cap. This would result in a down draft of the flue system. All of this is supposed to be planned out at the time of installation. The house yesterday looked great for venting."

"Is there anything else?" Bob then asks.

BTU Buddy says, "We haven't talked about combustion air for the furnace, but we'll have to hold that for another day."

**Capacity of Single-Wall Metal Pipe or Type B Asbestos Cement Vents**
**Serving a Single Draft Hood Equipped Appliance**

| Height H (ft) | Lateral L (ft) | Vent Diameter — D | | | | | | | |
|---|---|---|---|---|---|---|---|---|---|
| | | 3" | 4" | 5" | 6" | 7" | 8" | 10" | 12" |
| | | Maximum Appliance Input Rating in Thousands of Btu per Hour | | | | | | | |
| 6 | 0 | 39 | 70 | 116 | 170 | 232 | 312 | 500 | 750 |
| | 2 | 31 | 55 | 94 | 141 | 194 | 260 | 415 | 620 |
| | 5 | 28 | 51 | 88 | 128 | 177 | 242 | 390 | 600 |
| 8 | 0 | 42 | 76 | 126 | 185 | 252 | 340 | 542 | 815 |
| | 2 | 32 | 61 | 102 | 154 | 210 | 284 | 451 | 680 |
| | 5 | 29 | 56 | 95 | 141 | 194 | 264 | 430 | 648 |
| | 10 | 24* | 49 | 86 | 131 | 180 | 250 | 406 | 625 |
| 10 | 0 | 45 | 84 | 138 | 202 | 279 | 372 | 606 | 912 |
| | 2 | 35 | 67 | 111 | 168 | 233 | 311 | 505 | 760 |
| | 5 | 32 | 61 | 104 | 153 | 215 | 289 | 480 | 724 |
| | 10 | 27* | 54 | 94 | 143 | 200 | 274 | 455 | 700 |
| | 15 | NR | 46* | 84 | 130 | 186 | 258 | 432 | 666 |
| 15 | 0 | 49 | 91 | 151 | 223 | 312 | 420 | 684 | 1040 |
| | 2 | 39 | 72 | 122 | 186 | 260 | 350 | 570 | 865 |
| | 5 | 35* | 67 | 110 | 170 | 240 | 325 | 540 | 825 |
| | 10 | 30* | 58* | 103 | 158 | 223 | 308 | 514 | 795 |
| | 15 | NR | 50* | 93* | 144 | 207 | 291 | 488 | 760 |
| | 20 | NR | NR | 82* | 132* | 195 | 273 | 466 | 726 |
| 20 | 0 | 53* | 101 | 163 | 252 | 342 | 470 | 770 | 1190 |
| | 2 | 42* | 80 | 136 | 210 | 286 | 392 | 641 | 990 |
| | 5 | 38* | 74* | 123 | 192 | 264 | 364 | 610 | 945 |
| | 10 | 32* | 65* | 115* | 178 | 246 | 345 | 571 | 910 |
| | 15 | NR | 55* | 104* | 163 | 228 | 326 | 550 | 870 |
| | 20 | NR | NR | 91* | 149* | 241* | 306 | 525 | 832 |
| 30 | 0 | 56* | 108* | 183 | 276 | 384 | 529 | 878 | 1370 |
| | 2 | 44* | 84* | 148* | 230 | 320 | 441 | 730 | 1140 |
| | 5 | NR | 78* | 137* | 210 | 296 | 410 | 694 | 1080 |
| | 10 | NR | 68* | 125* | 196* | 274 | 388 | 656 | 1050 |
| | 15 | NR | NR | 113* | 177* | 258* | 366 | 625 | 1000 |
| | 20 | NR | NR | 99* | 163* | 240* | 344 | 596 | 960 |
| | 30 | NR | NR | NR | NR | 192* | 295* | 540 | 890 |
| 50 | 0 | NR | 120* | 210* | 310* | 443* | 590 | 980 | 1550 |
| | 2 | NR | 95* | 171* | 260* | 370* | 492 | 820 | 1290 |
| | 5 | NR | NR | 159* | 234* | 342* | 474 | 780 | 1230 |
| | 10 | NR | NR | 146* | 221* | 318* | 456* | 730 | 1190 |
| | 15 | NR | NR | NR | 200* | 292* | 407* | 705 | 1130 |
| | 20 | NR | NR | NR | 185* | 276* | 384* | 670* | 1080 |
| | 30 | NR | NR | NR | NR | 222* | 330* | 605* | 1010 |

*See Note 6

**FIGURE 57-4.** This vent sizing table came from the National Fuel Gas Code book and is used to size vent pipes. It takes into consideration the lateral (horizontal) vent matched to the vertical vent that terminates through the roof.

"What's combustion air?" Bob asks.

BTU Buddy explains, "It's the air that must be provided for any furnace for combustion."

Bob says, "I thought furnaces just use the air that's around the furnace."

"It's much more involved than that. Save that for next time," BTU Buddy says.

Bob says, "Well, I guess I'll have to. My brain is about full for now anyway."

# Combustion Air for a Gas Furnace

The winter season is just beginning and equipment problems are starting to surface. The dispatcher calls and tells Bob that yet another person is describing an unusual smell when the furnace runs for a long period of time.

Bob arrives and talks to the homeowner, who tells Bob, "The furnace has been running off and on for several days with no problems. But this morning it ran for a long time and was producing a strange odor."

Bob goes to the furnace room. The furnace is running, and he can smell aldehydes in the area. He knows that products of combustion contain aldehydes that are not poisonous, but are present in combustion gasses. He goes to the truck and gets a candle that he uses to see if the furnace draft is working. The furnace is back drafting slightly.

Bob is scratching his head, when BTU Buddy shows up. Bob says, "It's good to see you; you always seem to appear when I have a problem."

BTU Buddy says, "Looks like you have a draft problem. What are you going to do first?"

Bob explains, "I think I'll examine the flue to see if it's in good shape. I found a bird's nest in one the other day."

He gets a ladder and goes to the top of the house and removes the flue cap. The flue cap is built so that no birds can nest in it, but he removes it anyway. He can see by looking down the flue with a flashlight that there's no obstruction. He comes down to the furnace room and examines the actual flue connected to the furnace; it's double-wall pipe, as it should be. The flue system looks good. He asks BTU Buddy, "What do you think?"

BTU Buddy says, "I see the problem, but I want you to look around a little longer and see if you figure it out before I tell you."

Bob looks around again, when the furnace starts drafting like it should. The furnace room door is open to the rest of the house, and after a few minutes BTU Buddy says, "Close the door to the room and see what happens."

Bob closes the door and within a few minutes, the furnace stops drafting again. Bob looks around and discovers a rug covering a floor grill in front

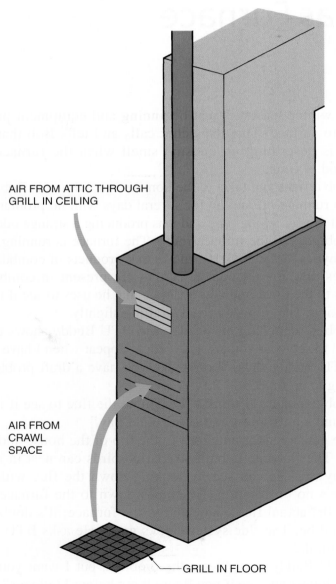

AIR FROM ATTIC THROUGH
GRILL IN CEILING

AIR FROM
CRAWL
SPACE

GRILL IN FLOOR

**FIGURE 58-1.** This illustration shows how the combustion air and dilution air reach this standard efficiency furnace.

of the furnace. When he removes the rug, a flow of cool air comes into the room and the furnace starts drafting, Figure 58-1. BTU Buddy says, "That was the problem. Somebody came into the room and felt the draft, and covered the fresh air intake, or makeup air, to the room. That grill must be left un-obstructed for the furnace to operate correctly.

"We've had several service calls concerning venting gas furnaces and have covered some of the basics. Let's look at some of the basics for combustion air for furnaces; it's just as important for a furnace to get the correct air supply to support combustion as it is to vent it correctly. A furnace must be supplied a specific amount of fresh air per cubic foot of gas burned. To burn 1 cubic foot of gas, approximately 1000 BTU for natural gas, requires 10 cubic feet of air for perfect combustion. Ten cubic feet of air contains 2 cubic feet of oxygen and 8 cubic feet of nitrogen and other gasses, combined with water vapor," Figure 58-2. "Air contains about 20% oxygen, so it takes 5 cubic feet of air to obtain 1 cubic foot of oxygen," Figure 58-3.

"These extra gasses just pass through the combustion process and must be exhausted through the flue system. But that isn't the end of the story. Perfect combustion is for laboratory studies and is not practical for field equipment.

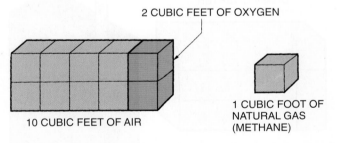

**FIGURE 58-2.** This illustration shows the combustion air requirements for perfect combustion.

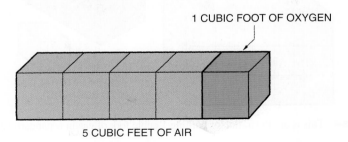

**FIGURE 58-3.** There is 1 cubic foot of oxygen for every 5 cubic feet of air.

Practical combustion and the regulations call for excess air to be furnished to the combustion process. Typically 50% excess air is used. So, the furnace must be provided with 15 cubic feet of air for each cubic foot of gas or 1000 BTU of input for natural gas. In addition to the 50% air for combustion, another 10 cubic feet of air and 50% excess air must be provided as dilution air for the venting process." Figure 58-4. "This other 50% goes up the draft diverter on standard furnaces to carry the products of combustion up the flue. Other gasses, such as propane, have tables for calculating air for them."

Bob exclaims, "There sure is a lot to this! Who's supposed to do all of this calculating?"

BTU Buddy explains, "The estimator is supposed to calculate this, but the service technician must be able to examine a service job and determine if the calculations are correct. The technician may be responsible for checking applications installed by other companies years after a furnace is installed."

Bob then asks, "How do you know what size grill is the right one?"

BTU Buddy says, "There are rules of thumb used for sizing and checking jobs in the field. The make-up air for the house we are in must come from

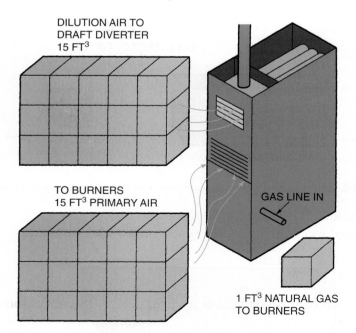

**FIGURE 58-4.** This is an illustration of burning 1 cubic foot of gas in a standard gas furnace.

two places. It's a one-story house built on a crawl space with an attic, and the combustion air sources are easy to deal with. There must be air for the burner that comes from near the floor, and air for venting that comes from up high. In this case, the burner air comes from the grill that was covered, and the air for venting comes from that vent in the ceiling that you haven't mentioned."

Bob looks up and says, "I thought I felt a breeze, but I didn't notice where it was coming from—and it's part of the system! These two grills are about the same; how should I check them to see if they're the correct size?"

BTU Buddy explains, "The National Fuel Gas Code calls for 1 square inch of free area for the dilution air and 1 square inch free area for the combustion air for each 1000 BTU per hour of input for the furnace. This is a 60,000 BTUH output furnace and it's of the 80% efficiency type, natural draft, so it's a 75,000 BTUH input furnace (60,000/.80 = 75,000 input). There should be two grills that each have a free area of 75 square inches."

Bob asks, "What do you mean by free area?"

BTU Buddy says, "Imagine that a 12 inch × 12 inch grill has an area of 144 square inches (12 × 12 = 144). That's grill area, but there are louvers inside the grill that take up room, obstruct the flow, or take up area. Most technicians multiply the area by 0.70 (for 70%) of the actual area to calculate the actual usable or free area. So the 144 square inch grill is actually only 100.8 square inches (144 × 0.70 = 100.8 free area)."

Bob asks, "Okay, how do we calculate the grill sizes for this furnace?"

BTU Buddy explains, "Let's use 12 inches for one side of the grill and calculate the other side. One reason to use 12 inches is that it will fit between rafters for the ceiling and floor joists that are in the floor. For larger furnaces, you can use 14 inches. The rafters and floor joists are on 16-inch centers, so a 14-inch grill will go between. The formula is X = furnace input in cubic feet per hour/common side × 0.70.

- X = the side we want to know
- furnace input in BTUH/1000 BTU/cubic foot = cubic feet per hour
- common side = 12 inches
- divided by 0.70 for free area calculation percentage

X = 75 cu ft/12 inches × 0.70
X = 8.93 inches

A 10-inch by 12-inch grill is a standard size. It will fit fine and have some excess capacity."

Bob said, "Boy, that's a lot to remember. Is there more?"

BTU Buddy says, "Yes, this is only for this type of structure," Figure 58-5. "There are structures on a slab, structures that are more than two stories, and structures that have flat roofs. All of these have a different plan. This service call is just to get you to start thinking. Now you should get a copy of the National Fuel Gas Code and look at it. The good thing about it is that there are pictures showing how the various types of structures are installed."

Bob asks, "Have these codes always been in force? I've seen many houses that I'm sure didn't have any combustion air or dilution air vents."

BTU Buddy then explains, "The codes were created in the early 1970s in many localities, around the time of the first energy shortage. Until then, homes weren't constructed as tightly as they are today. The windows and cracks around the structures allowed infiltration of air that was adequate for combustion air needs. With the energy shortage, energy prices started going up and people began to tighten down houses with storm windows and doors. Construction codes began calling for tighter homes, so suddenly there was a need to furnish planned, fixed ventilation for combustion air needs. Now is the time to talk to the homeowner and explain that these vents must not be covered for any reason."

FRESH AIR
FOR
COMBUSTION

**FIGURE 58-5.** This shows a structure that is like the structure in the article.

Bob explains the situation to the homeowner, who is happy for such a simple solution. Another problem solved.

While riding away, Bob says, "There sure seems to be a lot of air coming into that house for combustion, even though it's necessary. Is there no other way?"

BTU Buddy says, "Great question. Many newer furnaces have the combustion air pulled in through a pipe and exhausted through another pipe. The furnace and venting system can be much more efficient than the natural draft system in this house, but it's also much more expensive. Those systems have a combustion blower to create the movement of combustion gasses, rather than the natural draft system of this furnace."

Bob says, "I'm getting this, one step at a time. Thanks for the help."